ASTM INTERNATIONAL

SELECTED TECHNICAL PAPERS
STP1601

Editors: John J. Meyer and Josh Beakley

Concrete Pipe and Box Culverts

ASTM STOCK #STP1601
DOI: 10.1520/STP1601-EB

ASTM International, 100 Barr Harbor Drive, PO Box C700, West Conshohocken, PA 19428-2959
Printed in the U.S.A.

Library of Congress Cataloging-in-Publication Data

Names: Meyer, John, 1952- editor. | Beakley, Josh, 1967- editor. | ASTM International, sponsoring body.
Title: Concrete pipe and box culverts / editors, John Meyer, Josh Beakley.
Description: West Conshohocken, PA : ASTM International, [2017] | Series: Selected technical papers ; STP1601 | Papers presented at a symposium held December 7, 2016, in Orlando, Florida, USA sponsored by ASTM International Committee C13 on Concrete Pipe. | "ASTM Stock #STP1601." | Includes bibliographical references.
Identifiers: LCCN 2017019653 (print) | LCCN 2017019985 (ebook) | ISBN 9780803176461 (ebook) | ISBN 9780803176454 (pbk.)
Subjects: LCSH: Concrete culverts--Congresses.
Classification: LCC TE213 (ebook) | LCC TE213 .C66 2017 (print) | DDC 625.7/342--dc23
LC record available at https://lccn.loc.gov/2017019653

ISBN: 978-0-8031-7645-4

Peer Review Policy

Each paper published in this volume was evaluated by two peer reviewers and at least one editor. The authors addressed all of the reviewers' comments to the satisfaction of both the technical editor(s) and the ASTM International Committee on Publications.

The quality of the papers in this publication reflects not only the obvious efforts of the authors and the technical editor(s), but also the work of the peer reviewers. In keeping with long-standing publication practices, ASTM International maintains the anonymity of the peer reviewers. The ASTM International Committee on Publications acknowledges with appreciation their dedication and contribution of time and effort on behalf of ASTM International.

Citation of Papers

When citing papers from this publication, the appropriate citation includes the paper authors, "paper title," STP title, STP number, book editor(s), ASTM International, West Conshohocken, PA, year, page range, paper doi, listed in the footnote of the paper. A citation is provided on page one of each paper.

Printed in Bay Shore, NY
June, 2017

Foreword

THIS COMPILATION OF Selected Technical Papers, STP1601, *Concrete Pipe and Box Culverts*, contains peer-reviewed papers that were presented at a symposium held December 7, 2016, in Orlando, Florida, USA. The symposium was sponsored by ASTM International Committee C13 on Concrete Pipe.

Symposium Chairpersons and STP Editors:

John J. Meyer
Consultant
Wales, WI, USA

Josh Beakley
American Concrete Pipe Association
Irving, TX, USA

Contents

Overview

As we approach the end of the second decade of the twenty-first century, we must continue to move forward with an enlightened vision, building on the tremendous advancements made by the concrete pipe industry since its formation in the early 1900s.

This series of selected technical papers (STP) was published as the end-result of the December 2016 symposium on Concrete Pipe and Box Culverts, held in Orlando, Florida. The event was sponsored by ASTM Committee C13 on Concrete Pipe.

The objectives of this symposium were to present historical information on the evolution of specifications and manufacturing technology; describe new design and installation procedures; discuss innovative applications and uses; introduce new technologies for concrete pipe products; and to both discuss and determine the use of, and the need for, new ASTM standards for these products.

Concrete pipe products include circular pipe, box culverts, and manholes, along with all the other various shapes of pipe, and the innovative applications of precast concrete drainage devices.

The symposium met its objectives because of the countless hours dedicated to this undertaking by the authors/presenters. Not to be overlooked are the additional hours donated by those who performed peer reviews. These steps assure the international scientific and engineering community a quality publication.

Symposium Co-Chairmen and Editors

John J. Meyer, P.E.
Consultant
Wales, WI

Josh Beakley, P.E.
American Concrete Pipe Assoc.
Irving, TX

STP 1601, 2017 / available online at www.astm.org / doi: 10.1520/STP160120160123

George Hand II,[1] David Schnerch,[2] and Kimberly L. Spahn[3]

Lap Weld Strength of Reinforced Concrete Pipe Cages

Citation

Hand, G., Schnerch, D., and Spahn, K. L., "Lap Weld Strength of Reinforced Concrete Pipe Cages," *Concrete Pipe and Box Culverts, ASTM STP1601*, J. J. Meyer and J. Beakley, Eds., ASTM International, West Conshohocken, PA, 2017, pp. 1–17, http://dx.doi.org/10.1520/STP160120160123[4]

ABSTRACT

ASTM C76, *Standard Specification for Reinforced Concrete Culvert, Storm Drain, and Sewer Pipe*, and ASTM C1417, *Standard Specification for Manufacture of Reinforced Concrete Sewer, Storm Drain, and Culvert Pipe for Direct Design*, specify the requirements for reinforced concrete pipe, including the requirements for cage reinforcing welded lap splices. A discrepancy in the pull test requirement for welded lap splices existed between the 2008 versions of both specifications until ASTM C1417-11 was revised to mirror ASTM C76-08. ASTM C76-08 required pull tests of representative specimens to develop at least 50 % of the minimum specified tensile strength (or ultimate strength) of the steel. ASTM C1417-08 specified that pull tests of representative specimens develop no less than 0.9 times (or 90 %) the design yield strength of the circumferential. This discrepancy raised questions when ASTM C1417-08 was revised as to how the required lap splice strength of 90 % of the yield strength was established for ASTM C1417-08 and concern that this might be a more appropriate requirement. Therefore, the concrete pipe industry produced and tested 24-in. and 36-in. diameter pipe in which the welded lap splices did not meet the ASTM C76 requirement of 50 % ultimate strength in order to demonstrate that this requirement does not affect the final three-edge bearing product test results.

Manuscript received September 6, 2016; accepted for publication October 18, 2016.

[1]Oldcastle Precast, 1920 12th St., Folsom, NJ 08037

[2]Wiss, Janney, Elstner Associates, Inc., 311 Summer St., Boston, MA 02210

[3]American Concrete Pipe Association, 8445 Freeport Pkwy., Suite 350, Irving, TX 75063

[4]ASTM Symposium on *Concrete Pipe and Box Culverts* on December 7, 2016 in Orlando, Florida.

Keywords
welding, reinforcement, concrete pipe, cages, manufacturing, welded wire reinforcement

Introduction

Reinforced concrete pipe (RCP) is a composite material typically manufactured from two major structural components—concrete and steel reinforcing cages, as shown in **Fig. 1**. Pipe walls are subjected to bending and shear stresses during handling, transport, installation, and in-service; therefore, the function of the cage is to resist the tensile stress in the pipe wall at the locations shown in **Fig. 2**. The pipe cage also supports freshly cast concrete during the manufacturing process prior to the concrete hardening because the external mold is removed almost immediately after casting.

The reinforcement in concrete pipe provides a significant amount of strength and ductility. "Reinforced concrete pipe, like other reinforced concrete structures, is designed to crack. RCP design accommodates the high compressive strength of concrete and the high tensile strength of steel. As load on the pipe increases, and the tensile strength of the concrete is exceeded, cracks will form as the tensile load is transferred to the steel" [1]. **Fig. 3** shows the ductility provided by the reinforcing steel. The area shown in red indicates the ductility in pipe without

FIG. 1 Manufacturing of RCP.

FIG. 2 Pipe bending stress zones under load.

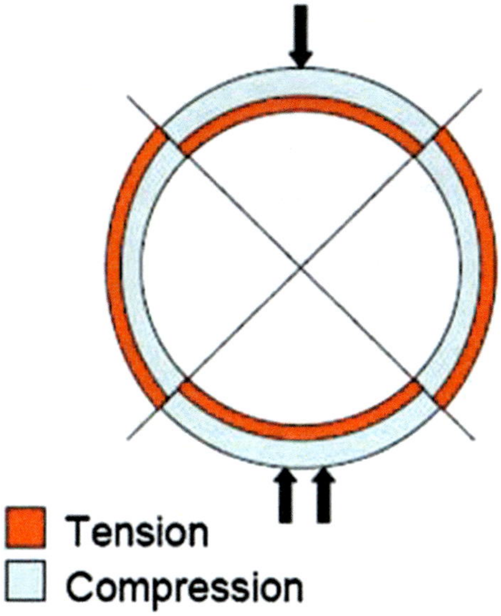

reinforcement, and the area shown in blue indicates the continued ductility of reinforced pipe.

The three-edge bearing test (3EB), as shown in **Fig. 4a–c** and specified in ASTM C497-16a, *Standard Test Methods for Concrete Pipe, Manhole Sections, or Tile* [3], is used to test the final composite of concrete and steel reinforcement, including welds at the lap splice, to determine the strength of the product.

Types of Cages

Reinforcing cages may be produced from smooth or deformed welded-wire reinforcement (WWR) or helically wound from cold-drawn wire. Both types of cages have longitudinal wires and circumferential wires. However, only cages produced from WWR have a splice or lapped splice. WWR is prefabricated from high-strength, cold-drawn, or cold-rolled wires. Each wire intersection is resistance-welded by automatic welders and is governed by ASTM A1064-16, *Standard Specification for Carbon-Steel Wire and Welded Wire Reinforcement, Plain and Deformed, for Concrete* [4].

Helically wound cages wind a single strand of wire continuously into a spiral-type shape around multiple longitudinal wires as shown in **Fig. 5**. The longitudinal wires serve to keep the wound circumferential wire in place during the production of the pipe until the concrete cures. Therefore, as shown in ASTM C76-15a, *Standard Specification for Reinforced Concrete Culvert, Storm Drain, and Sewer Pipe* (Section 6.5), a helically wound cage is exempt from testing the weld for shear

FIG. 3 Typical load-deflection curve of pipe in three-edge bearing test [2].

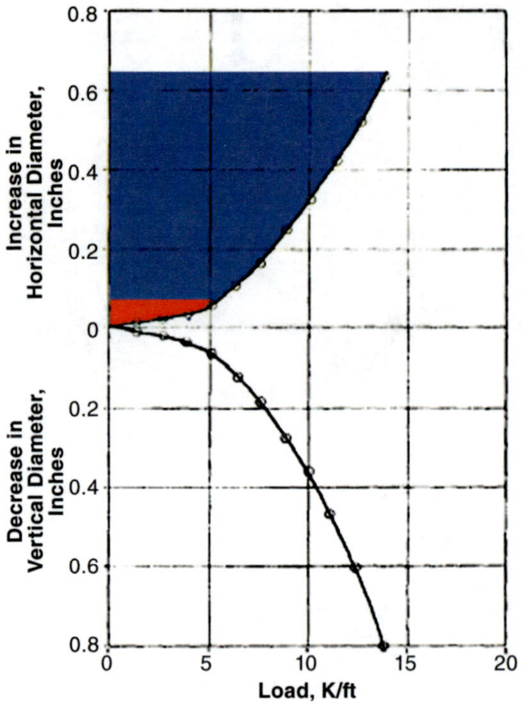

between the longitudinal wire and the circumferential wound wire because that weld is nonstructural [5].

Unlike helically wound cages, cages made from WWR must develop a means of transferring tensile stresses at the intersection of the two ends of the WWR mat or roll in order to achieve structural continuity. Cage rollers are used to bend flat sheets or rolls of WWR to the correct diameter and to cut the appropriate length, as shown in **Fig. 6**. Once the cage is cut to length, the circumferential wires are spliced and either have a longer lap without a weld or are welded with a short 2-in. lap. Smooth wire reinforcement relies on the bond to concrete by the mechanical anchorage at each longitudinal wire intersection. Therefore, according to ASTM C76-15a, if the smooth wire ends are not being welded, the manufacturer must lap the two ends by 40 wire diameters and the lap shall contain a longitudinal wire. Deformed reinforcement gains additional mechanical anchorage from the deformations and therefore is required to lap only a length of 20 wire diameters and shall contain a longitudinal wire. If the ends are welded, ASTM C76-15a, Section 8.1.8.1 (which is now superseded by Active Standard ASTM C76), states that there still must be a 2-in. lap and the weld must develop at least 50 % of the minimum specified tensile (or ultimate) strength of the steel as shown in **Fig. 7**.

FIG. 4 (a) Three-edge-bearing (3EB) test, (b) graphical representation of 3EB test, and (c) example of 3EB test being performed—measurement of 0.0-in. crack after reinforcement has accepted load.

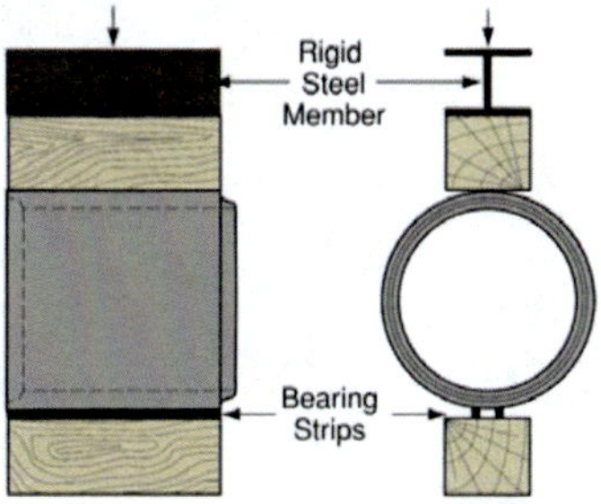

(a)

(b)

(c)

History of Weld Strength in RCP

As stated earlier, ASTM C76 was first to require a welded lap strength of 50 % of the ultimate strength for WWR when it is welded with a lap length of less than 40 or 20 wire diameters for smooth or deformed wires, respectively. This is believed to have been established when RCP utilized mild steel that had a 40 ksi yield strength and an 80 ksi ultimate strength. This meant that welded WWR was tested to ensure it would pass the yield strength of the wire before the weld failed. Production practices today utilize 60 ksi to 70 ksi yield steel, but the ultimate strength is typically 80 ksi to 100+ ksi. Although the yield strength has moved closer to the ultimate

FIG. 5 Helically wound cage production.

strength, the industry still tests the weld to 50 % of the ultimate strength with no known problems with products in the field.

However, in 2008, ASTM C1417-08 (which is now superseded by Active Standard ASTM C1417), *Standard Specification for Manufacture of Reinforced Concrete*

FIG. 6 WWR cage rolling machine.

FIG. 7 ASTM C76-15a, Section 8.1.8.1 (which is now superseded by Active Standard ASTM C76).

8.1.8.1 When splices are welded and are not lapped to the minimum requirements above, there shall be a minimum lap of 2 in. and a weld of sufficient length such that pull test of representative specimens shall develop at least 50% of the minimum specified tensile strength of the steel. For butt-welded splices in bars or wire, permitted only with helically wound cages, pull tests of representative specimens shall develop at least 75% of the minimum specified tensile strength of the steel.

Sewer, Storm Drain, and Culvert Pipe for Direct Design [6], added the requirement to Section 10.4 that WWR circumferential wires lapped by welding must still contain the minimum 2-in. lap, but the weld must develop no less than 90 % of the yield strength, as shown in Fig. 8. If a manufacturer is using WWR with a 70 ksi yield strength, the weld pull test in ASTM C1417-08 requires that the wire pass a 63 ksi weld pull test, which is 57 % higher than the ASTM C76-08 requirement. This language was based on research performed by Spiekerman [7]. The research tested WWR with a 1-in. weld and 2-in. lap encased in a concrete cylinder. Spiekerman's research confirmed, as shown in Fig. 9, that this configuration could achieve 90 % of its yield strength. However, the research did not go as far as testing the configuration in the final RCP product. Therefore, in 2014, ASTM C1417-14 changed its specification to match ASTM C76-08 because 90 % of yield strength was overly conservative for the performance of the pipe; to date, there are still no known failures of RCP due to the welds of the circumferential wire lap splice.

Testing of Weld Strength in RCP

After changing ASTM C1417-14 to mirror ASTM C76-08 with a 50 % ultimate weld pull test, Pennsylvania raised concern about the validity of that change. Therefore, in 2015, the American Concrete Pipe Association (ACPA) partnered with the Pennsylvania Department of Transportation (PennDOT) and performed a study on whether or not a weld made to 50 % of ultimate was sufficient. Three different manufacturers (Northern Concrete Pipe: Bay City, MI; Oldcastle Precast: Croydon, PA; and Cretex, Inc.: Elk River, MN) participated in the study and intentionally

FIG. 8 ASTM C1417-08, Section 10.4 (which is now superseded by Active Standard ASTM C1417).

10.4 *Lapped Splices of Circumferential Reinforcement:*

10.4.1 Where lapped circumferentials are spliced by welding, they shall be lapped no less than 2 in. Pull tests of representative specimens shall develop no less than 0.9 times design yield strength of the circumferential.

FIG. 9 Cylinder with welded lap splice used to determine 90 % yield requirement in ASTM C1417-08.

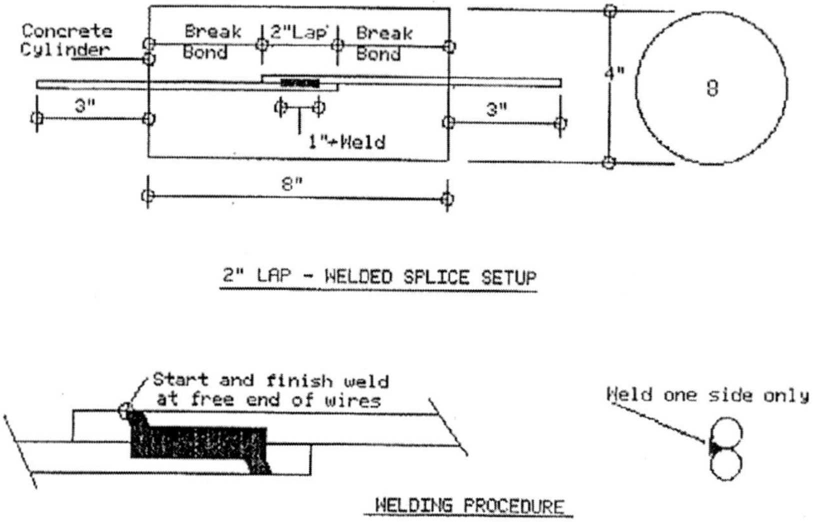

produced welds less than the ASTM C76-08 requirement. The test protocol to create the worst case scenario was as follows:

1. Each producer was to complete the welds using standard welding procedures.
2. All three manufacturers produced their pipe as closely as possible to the PennDOT concrete strengths and reinforcing requirements, as shown in **Table 1**. The design and manufacturing criteria that were used for the study are shown in **Table 1**. There are some differences between the PennDOT specification and the ASTM C76-15a specification. As fill heights increase, the significance of flexural behavior diminishes and requirements for crack control dominate the design. In the case of PennDOT, there is more steel (greater than 30 % more cross-sectional area) because a soil unit weight of 140 pcf is utilized instead of 120 pcf, as is commonly used for RCP design. For relatively shallow fill heights where flexure behavior controls, PennDOT uses a more conservative phi factor of 0.9 for the Type A pipe and a phi factor of 0.95 for the Type B pipe, where ASTM C76-15a uses a phi factor of 1.0. For designs where crack control governs the design, PennDOT uses a maximum 0.007-in. crack width instead of the common 0.01-in. commonly used in ASTM C76-15a.
3. Each manufacturer was to produce a pipe with a lap splice of sufficient length to meet the requirements of ASTM C76-15a with no weld at all and another pipe with a 2-in. lap and a 1/4 in. long tack weld, as shown in **Fig. 10**. However, both Northern Concrete Pipe and Oldcastle Precast utilized a tack weld less than $^{1}/_{2}$ in. and still met the other parameters of the study.

TABLE 1 4,000 psi concrete strength (PennDOT fill height 7–10 ft versus ASTM C76-15a, Class II).

Diameter	PennDOT Wall (in.)	C76 Wall (in.)	PennDOT Steel Area (sq. in./ft.)		C76 Steel Area (sq. in./ft.)		PennDOT Proof Test Load (d-load) = .007″ crack	C76 D-Load = 0.01″ crack
			Inner Cage	Outer Cage	Inner Cage	Outer Cage		
24″	3	3	0.1		0.07		886 lbs/ft.	1,000 lbs/ft.
	3.75	3.75	0.08		0.07		886 lbs/ft.	1,000 lbs/ft.
36″	4	4	0.16		0.2		881 lbs/ft.	1,000 lbs/ft.
			0.12	0.07	0.12	0.07		
	4.75	4.75	0.14		0.16		881 lbs/ft.	1,000 lbs/ft.
			0.1	0.07	0.07	0.07		

4. All pipe was 3EB-tested within three days after production in order to address concerns with testing pipe after the concrete compressive strength rises above 4,000 psi. This concrete compressive strength is the minimum required by ASTM C76-15a for the sizes and class of pipe used in the investigation. However, concrete strength greater than 4,000 psi would not affect the results of the weld strength tests because, in order for stress to develop in the weld greater than about 3,400 psi, the concrete must be cracked independent of the concrete strength. For example, if the concrete strength is 4,000 psi, typical cracking strength is about 475 psi and, for 6,000 psi, about 580 psi. The corresponding

FIG. 10 Lap and weld configuration for ACPA study.

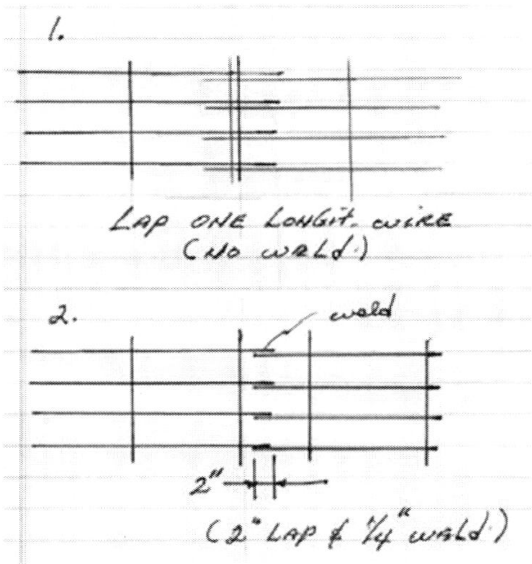

TABLE 2 Northern Concrete pipe—24-in. pipe properties produced.

	PennDOT Requirement	Actual Tested
Diameter	24″	24″
Wall Thickness	3″	3″
Area of Steel	0.10 sq. in./ft.	0.10 sq. in./ft.
Concrete Strength	4,000 psi	5,518 psi

stress in the steel would be about 3,400 psi. Until the concrete has cracked, the strain in the steel and concrete is the same. In order to be assured the concrete is cracked (the initial cracking may not be visible), the minimum D-load strength should not be less than 1,350 lb/ft/ft.

5. All cages were marked at the weld or lap. Then pipe was oriented in the 3EB test with the lap or weld to coincide with the position of maximum tensile stress. The critical position is at the invert for pipe using an inner or single cage and at the spring line for the outer cage.
6. After recording the results of the 3EB test for the 0.01-in. crack, all pipe were taken to ultimate failure and the results recorded.

Test Results of Weld Strength in RCP

NORTHERN CONCRETE PIPE, BAY CITY, MI

Table 2 shows the parameters specified to manufacture pipe in accordance with the PennDOT specification and the parameters to which the pipe was actually manufactured.

The test protocol, as described in the previous section, did not require pull tests of the weld; however, the pull tests performed by Northern Concrete Pipe and reported in **Table 3** demonstrate that representative welds intentionally did not meet the requirement to achieve 50 % of the ultimate steel strength (80 ksi). However, as shown in **Table 4**, the 3EB strength requirement is being achieved. All welds tested were cut from reinforcing cages used in production of pipe and tested for the

TABLE 3 Northern Concrete pipe—weld pull test.

Wire Diameter	Tensile Strength (psi)	Description
0.179	35,040	Welded at longitudinal wire 1/4″ Tack
0.178	26,097	Welded at center of circumferential wire lap
0.179	28,750	Welded at center of circumferential wire lap
0.179	35,159	Welded at end away from circumferential wire
0.177	63,390	3/4″ weld (not used in PennDOT test)

FIG. 11 Northern Concrete Pipe cage: (a) 2-in. lap (b) less than ½-in. weld.

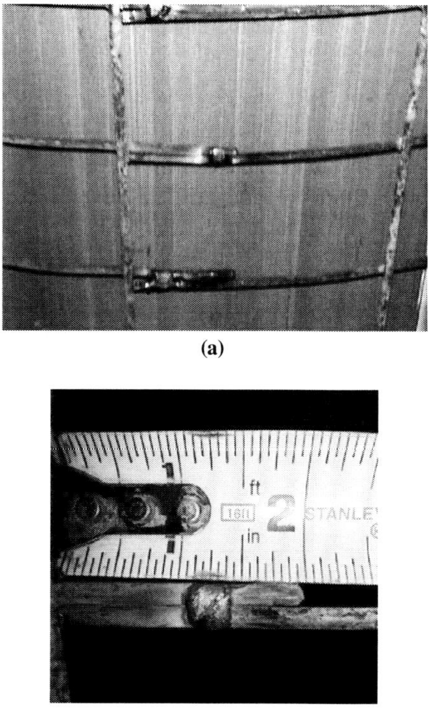

(a)

(b)

investigation. The uppermost circumferential wire from each cage was cut to be tested for pull strength. **Fig. 11a** and **b** shows the welds produced by Northern Concrete Pipe for the study.

As shown in **Table 4**, the tested pipe met both the required ASTM C76-15a, Class II, D-load of 1,000 lb/ft/ft for the 0.01-in. crack and 1,500 lb/ft/ft for the ultimate load as well as the load required to meet the PennDOT 0.007-in. crack of 886 lb/ft/ft. Some concern was raised that the 24-in. welded pipe's 0.01-in. D-load and

TABLE 4 Northern Concrete Pipe three-edge bearing results.

Type of Lap	Lap Location	0.007″ Crack (lbs/ft/ft)	0.01″ Crack (lbs/ft/ft)	Ultimate Load (lbs/ft/ft)
2″ Lap/ < ½″ Weld	Invert	1,982	1,982	2,104
	Crown	1,890	1,890	2,012
Tied Full Lap	Invert	2,012	2,012	2,408
	Crown	1,890	1,890	2,561

FIG. 12 Weld failure at ultimate load.

ultimate load were less than 10 % apart. However, all strengths met the loading requirements, and the testing crew took pictures of the welds popping, as shown in Fig. 12. So this proves that the reinforcement did take the load prior to the welds failing. There are no limitations in ASTM C76 on the separation of the 0.01-in. crack load and the ultimate load, only that the pipe must meet both parameters. Note, the 0.007-in. crack and 0.01-in. crack are reported at the same load. This is due to the small difference in physical size of the two crack dimensions. Once the pipe cracked, the measurement of the crack was at 0.01 in.; therefore, the 0.007-in. crack met the same load.

OLDCASTLE PRECAST, CROYDON, PA

Because the study was a partnership between PennDOT and the ACPA, it was requested that the manufacturer located locally in Pennsylvania allow PennDOT to take samples of the welds to pull test at their own facility. Oldcastle produced their

TABLE 5 Results of pull test conducted by PennDOT.

Tensile Strength (psi)	Pass/Fail PennDOT	Pass/Fail C76
31,320	Fail	Fail
46,080	Pass	Pass
38,960	Fail	Fail
39, 240	Fail	Fail
41,920	Fail	Pass
38,600	Fail	Fail
29,800	Fail	Fail
39,440	Fail	Fail
AVG = 33,265	Fail	Fail

TABLE 6 Oldcastle Precast three-edge bearing test results.

	PennDOT Requirement	Actual Tested
Diameter	24″	24″
Wall Thickness	3″	3″
Area of Steel	0.10 sq. in./ft.	0.10 sq. in./ft.
Concrete Strength	4,000 psi	Set 1: 4,180 psi
		Set 2: 5,176 psi

pipe for the study using a wire diameter of 0.1785 in. with approximately 0.5-in. welds. The wire had an 80/65 ksi ultimate/yield strength, respectively. PennDOT requires the wire to pull test to 70 % of yield, which in this case is 45,500 psi, while ASTM C76's requirement of 50 % of ultimate should result in a test of 40,000 psi.

PennDOT selected the best welds from the cages; therefore, the welds left on the product may have had slightly lower weld strengths creating more conservative results with regard to the need for higher weld strengths in the 3EB test. The results of the pull test are shown in **Table 5**, and most of the welds intentionally have failed to meet either PennDOT's or ASTM C76's requirements. **Table 6** indicates that, even with welds that do not meet the specification, the final product passes all three D-load criteria with the test being stopped after reaching 15 % over the D-ultimate load.

Table 7 shows the parameters specified to manufacture pipe in accordance with the PennDOT specification and the parameters to which the pipe was actually manufactured by Oldcastle Precast. **Fig. 13** demonstrates how Oldcastle Precast marked where their weld or lap was located in order to ensure it was placed in the critical zone during the 3EB test.

CRETEX, INC., ELK RIVER, MN

Cretex produced a 36-in. Class III pipe. **Table 8** shows the parameters specified to manufacture pipe in accordance with the PennDOT specification and the parameters to which the pipe was actually manufactured by Cretex. **Fig. 14** shows the cage

TABLE 7 Oldcastle Precast—24-in. pipe properties produced.

Type of Lap	0.007″ Crack (lbs/ft/ft)	0.01″ Crack	Ultimate (lbs/ft/ft)
Set 1			
2″ Lap/<1/2″ Weld	1,437	1,437	Stopped at 0.01″+15 %
Tied Full Lap	1,312	1,312	Stopped at 0.01″+15 %
Set 2			
2″ Lap/<1/2″ Weld	1,469	1,469	Stopped at 0.01″+15 %
Tied Full Lap	1,250	1,250	Stopped at 1,250 lb/ft/ft w/< 0.007″

FIG. 13 Oldcastle Precast pipe marked and tested (a) finished test pipe with lap marked, (b) cage marked, and (c) pipe with lap at six o'clock in 3EB machine.

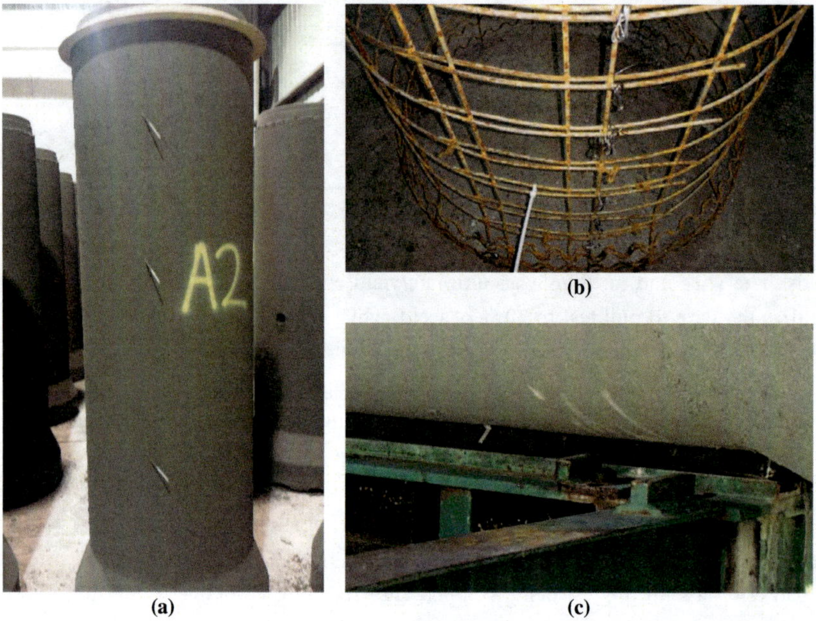

welded every circumferential; however, Cretex tested pipe also welded every *other* circumferential. **Table 9** provides the 3EB test results. The test load required to produce the 0.007-in. crack for a Class III pipe is 915 lb/ft/ft, and the ASTM C76 requirements for a 0.01-in. crack and ultimate load are 1,350 lb/ft/ft and 2,000 lb/ft/ft, respectively.

Conclusion of Test Results of Weld Strength in RCP

Due to the clear-cut results of all pipe tested passing not only the PennDOT specification but also the ASTM C76-15a specification with welds that did not meet

TABLE 8 Cretex—36-in. pipe properties produced.

	PennDOT Requirement	Actual Tested
Diameter	36″	36″
Wall Thickness	4″	4″
Area of Steel (inner/outer)	0.18/.12 sq. in./ft.	0.18/.10 sq. in./ft.
Concrete Strength	4,000 psi	5,179/5,312 psi

FIG. 14 Cretex welded cage.

specifications, it was determined that the weld shear currently required in ASTM C76-15a was sufficient. However, during the study, it was agreed that neither ASTM C76, ASTM C1417, nor ASTM C497 gives a standardized test procedure for the pull test required by ASTM C76 and ASTM C1417. Therefore, at this writing, ASTM Subcommittee C13.09 on Testing Procedures has balloted a new section in ASTM C497 to standardize that test.

Pull Test Procedure in ASTM C497 of WWR Weld in RCP

"This pull test method is proposed to cover procedures for the mechanical pull (tensile) testing of butt welded wire and welded wire reinforcement used for reinforced rigid concrete pipe and precast products. A representative specimen of the welded lap splice is tested in a machine designed to apply tension along the longitudinal

TABLE 9 Cretex three-edge bearing test results.

Type of Lap	First Crack	0.007″ Crack (lbs/ft/ft)	0.01″ Crack (lbs/ft/ft)	Ultimate Load (lbs/ft/ft)
1/4″ Weld Every Circumferential & 2″ lap	1,229	1,265	1,354	2,366
	1,258	1,327.50	1,370	2,225
1/4″ Weld Every other longitudinal & 2″ lap	1,285	1,347	1,433	2,330
	1,125	1,250	1,350	2,071
Normal lap	1,579	1,684	1,712	2,644
	1,458	1,541	1,583	2,975

FIG. 15 Weld pull test at PennDOT Laboratories: (a) Lap Weld 1 as cut, (b) Lap Weld 2 in vise, (c) Lap Weld 3 straightened, and (d) Lap Weld 4 ready to test.

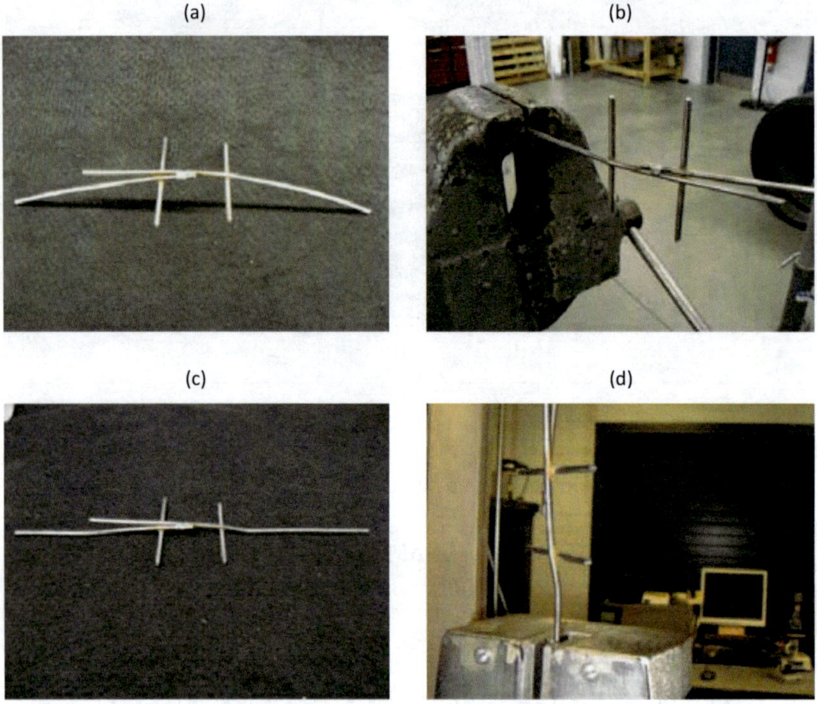

axis of the welded test specimen to determine the pull strength of the weld. The mechanical tests herein described are used to determine minimum tensile properties of the weld required in the product manufacturing specifications where indicated" [3]. **Fig. 15** shows the weld pull test being performed at the PennDOT Laboratories that is also being balloted in ASTM C497.

Conclusions

Although the ASTM C76 weld requirement of a 50 % ultimate wire strength has served the RCP industry for many years, the discovery of how a weld of a lap splice for a cage made from WWR that does not meet the 50 % minimum contributes to the final strength of the product was informative. The study was beneficial to industry and owners in order to prove a 90 % weld was not necessary and that the change in ASTM C1417 to mirror ASTM C76 was warranted. Additional benefit was realized that a standardized test procedure should be developed for the weld pull test. In the future, the industry may investigate whether or not a lower than 50 % ultimate weld strength requirement is sufficient.

ACKNOWLEDGMENTS

The authors acknowledge Northern Concrete Pipe, Inc., Oldcastle Precast, and Cretex, Inc., for their contribution to the ACPA/PennDOT weld of splices in WWR study.

References

[1] ACPA, *Cracks in Installed Reinforced Concrete Pipe*, American Concrete Pipe Association, Irving, TX, 2007, p. 1.

[2] Heger, F. and Liepins, A., "Stiffness of Flexurally Cracked Reinforced Concrete Pipe," *ACI J.*, Vol. 82, No. 3, 1985, pp. 331–342.

[3] ASTM C497, *Standard Test Methods for Concrete Pipe, Manhole Sections, or Tile*, ASTM International, West Conshohocken, PA, 2016, www.astm.org

[4] ASTM A1064-16, *Standard Specification for Carbon-Steel Wire and Welded Wire Reinforcement, Plain and Deformed, for Concrete*, ASTM International, West Conshohocken, PA, 2016, www.astm.org

[5] ASTM C76-16, *Standard Specification for Reinforced Concrete Culvert, Storm Drain, and Sewer Pipe*, ASTM International, West Conshohocken, PA, 2016, www.astm.org

[6] ASTM C1417-15, *Standard Specification for Manufacture of Reinforced Concrete Sewer, Storm Drain, and Culvert Pipe for Direct Design*, ASTM International, West Conshohocken, PA, 2015, www.astm.org

[7] Spiekerman, B., "WeldT," California Concrete Pipe Corp., Stockton, CA, 1988.

STP 1601, 2017 / available online at www.astm.org / doi: 10.1520/STP160120160118

Eric Carleton,[1] Steve Hiner,[2] and John Kurdziel[3]

The History and Application of the Three-Edge Bearing Test for Concrete Pipe

Citation

Carleton, E., Hiner, S., and Kurdziel, J., "The History and Application of the Three-Edge Bearing Test for Concrete Pipe," *Concrete Pipe and Box Culverts, ASTM STP1601*, J. J. Meyer and J. Beakley, Eds., ASTM International, West Conshohocken, PA, 2017, pp. 18–27, http://dx.doi.org/10.1520/STP160120160118[4]

ABSTRACT

The three-edge bearing test is one of the only direct test methods used for the evaluation of a finished concrete product. Many tests exist for the evaluation of the components of reinforced concrete structures and pavements, but none routinely evaluate the performance of the product with the applied loads on the finished product. This makes the three-edge bearing test one of the most unique structural evaluation tests in existence in the engineering and construction fields. The three-edge bearing test is approaching its 100-year anniversary and this paper presents the reasons for its initial development and its use in the present day. Technology has advanced considerably in the past century with finite element modeling, remote sensors, and computerized construction equipment, yet the three-edge bearing test is still the linchpin in the evaluation of reinforced concrete pipe design and installation. The three-edge bearing test is also positioned to be a key component in the assessments of future composite reinforced concrete pipe products, so although it may be nearing the century mark, this test is well-positioned to move into its next century. Understanding its past is, therefore, critical in evaluating the future of this key test method for reinforced concrete pipe.

Keywords

strength test, three-edge bearing, reinforced concrete pipe

Manuscript received August 31, 2016; accepted for publication November 9, 2016.

[1]Advanced Drainage Systems, Inc., 3450 River Forest Dr., Fort Wayne, IN 46805

[2]National Precast Concrete Association, 1320 City Center Dr., Carmel, IN 46032

[3]Rinker Materials, 18702 Arcaro Glen Ct., Houston, TX 77346

[4]ASTM Symposium on *Concrete Pipe and Box Culverts* on December 7, 2016 in Orlando, Florida.

Introduction

Anytime one does a summary of the historical development of any process, design, or significant event, the author(s) too often insert their own analysis and perspective upon it. By doing so, much of the historical evolution and thought process of the original developers is lost and the reader is left only with the interpretation of the subject matter by the author(s). The authors of this paper are no different as we are only human and, as such, are guilty of these same traits. In order to reduce the impact of this author bias, however, as much of the original words and quotes from the original developers of the three-edge bearing test will be used and referenced so the readers may assess the significance of the content and meaning independently from the interpretations of these authors.

The Early Years

The development of the three-edge bearing test, as with most tests, was born out of necessity and not from a sequence of logical analytical design methodology. The initial pipe load test [1] was associated with verifying the strength of drain tile, but these tests were more to determine and study the variability associated with drain tile strengths based on different manufacturing techniques and conditions. In this early study, researchers were not trying to associate the results with minimum requirements for field applications but only to get a handle on why there was so much variability in pipe strengths.

The most notable study occurred more than 100 years ago when it was noted that concrete drain tile was cracking after installation. The reasons were not clearly known at the time, so a very detailed study was conducted by Marston and Anderson [2] at Iowa State College to determine why these problems were occurring and how to design concrete drain tile to prevent these field issues. Their original assessment clearly indicates both the dearth of any design understanding and the lack of how one can properly assess the acceptability of a concrete pipe by just visual inspection.

> Engineers and inspectors simply give the pipe an external examination, and where there are no serious defects visible, try to determine by intuition whether they will carry safely the loads which must rest upon them. In many cases rejected pipe have been proven by tests to be stronger and better than accepted pipe for the same lot. In many cases, the sincerest efforts of both manufacturers and engineers have failed to exclude pipe which afterwards cracked in the ditch [2].

As the pipe diameters increased in diameter, the number of problems occurring in the field rose. These "failures of large drain tile by cracking in the ditches" were becoming extremely common and were deemed to be of serious concern with proper design and testing well overdue.

> The manufacture and use of tile and sewer pipe are of very great pecuniary importance. Moreover, the failure of agricultural drains may ruin the farmer's crops, and the failure of a sewer may endanger the health of a neighborhood. Considering the importance of the subject, and remembering that sewer pipe of fairly large diameters have been in extensive use for generations, it would certainly seem that standard methods for testing sewer pipe and drain tile should have been adopted and brought into general use long since [2].

As is common today, Marston and Anderson classified the failures in two cases: cracking that developed during construction and the second, "drain tile supposed to be all right are found to be cracked after a considerable time has elapsed since construction" [2]. They rightfully acknowledge that the installation cracking was a result of poor handling and construction techniques. The second condition was the result of dead and live loads based on the type of installation used for pipe embedment.

The majority of this first report went on to define how loads are distributed to a pipe in a trench. It was at this time, however, that Marston and Anderson did their first test to correlate the in-field loads to the pipe strength based on the manufacturer's own production process. They evaluated various pipe diameters with various wall thicknesses and mix proportions and assessed their applied loads, bearing strengths, and modulus of rupture. As one would expect with no standardization and multiple manufacturers, the resulting data were all over the place, but this represented the first assembled test data for concrete pipe, albeit all nonreinforced.

The first rudimentary testing machine was used to test a 36-in. cement drain tile. This apparatus was known as the Ames Standard Homemade Testing Machine.

It was not until four years later that Marston, along with Schlick and Clemmer, looked at standardizing the test protocol and manufacturing processes for concrete pipe [3]. Based on the content of the material contained in their report, however, it is believed that the first ASTM standard for concrete drain tile was created prior to the release of this work, essentially placing the origin of the preliminary load bearing test in the early 1910s. One of the precursors to the three-edge bearing test was the sand-bearing test, as indicated in **Fig. 1**.

Although the original efforts initiated by the Iowa State College researchers were to address drain tile issues, it was the more critical larger diameter, sewer pipe applications that drove the need for detailed design and testing protocols. The importance of these installations and the thrust of their goal are clearly indicated.

> In pipe sewer construction, however, engineers have been obliged to use rule of thumb methods to provide for the structural stability of constructions which have cost many hundreds of millions of dollars and which are vital to the health of hundreds of millions of people [3].

This section was followed by a simple statement by these Iowa professors that essentially summarizes what all engineering pipe design should be fundamentally

FIG. 1 Sand-bearing test method.

based upon, not just that for concrete pipe. Whatever doubt there might be about what is correct or appropriate, this should be the litmus test: "Rational methods of design of pipe sewers as to the structural strength should be substituted at once for the rules of thumb which have been used heretofore" [3].

When assessing the critical components for concrete pipe design, two key parameters were identified:

1. Determine the loads to be carried.
2. Prescribe the use of structures of such definitive known strengths as are considered amply sufficient to carry the loads safely under all contingencies.

The thought process that followed clearly defined not just the need but the means of combining those field-attributed loads with the means for providing a designed product that could resist said loads.

> Up to the present time, no standard method has been adopted for testing the supporting strength of sewer pipe. In this particular, sewerage engineering is behind drainage engineering, for the American Society for Testing and Materials (ASTM) adopted a standard method three years ago for testing the "ordinary supporting strength" of drain tile, so that all drainage engineers can now obtain test results which are comparable. Furthermore, drainage engineers are able to ascertain in advance, by the standard strength test, whether there is any danger that the pipe to be used will crack in any particular ditch; and are

> able to prevent [the] danger of cracking by proper specifications of the "ordinary supporting strengths" of sample pipe under the standard test [3].
>
> The main points that must be determined in devising a satisfactory standard test of the "ordinary supporting strength" of sewer pipe are:
> 1. What length of hub-spigot sewer pipe shall be used in calculating the "ordinary supporting strength" (per unit length) from the test cracking load of the whole pipe?
> 2. What formulas and what sewer pipe dimensions shall be used in calculating the modulus of rupture of the pipe?
> 3. What bearings shall be used in applying the loads in making tests of the "ordinary supporting strengths" of sewer pipe?
>
> The hub (expanded bell) and the barrel of the pipe are connected rigidity so that they must break together in all laboratory tests of the "ordinary supporting strength," whether the bell is loaded or not. It would seem that the hub must increase the cracking load to some extent, even when only the barrel is loaded in tests made with the "two point" and "three point" bearings. After a careful study of the whole subject and of the detailed results of the comparative tests whose results are given in Tables V to X, inclusive, below we have reached the conclusion that: *The net inside length of sewer-pipe from the bottom of the hub-socket to the extremity of the spigot-end should be used as the divisor in calculating the "ordinary supporting strength" (per unit length) from the total test cracking load on the whole pipe* [3].

It should be noted the aforementioned tests summarized in Tables V through X represent 380 sand-bearing tests on the diameters of 12-in., 18-in., and 24-in. pipe and 55 three-point bearing tests on 6-in., 12-in., and 18-in. pipe. The initial equation developed for the structural evaluation may have been lost or forgotten over the years, but based on all their original testing, the following formula was prepared to assess the bearing strength for concrete sewer pipe.

> The formulas for calculating the modulus of rupture of sewer pipe from the "ordinary supporting strength" should be the same as already adopted by the American Society for Testing Materials for drain tile, as follows:
>
> $$M = 0.20\ r\ (W/12) \tag{1}$$
>
> $$F = 6M/t^2 \tag{2}$$
>
> where:
>
> M = maximum bending moment in the pipe wall of the barrel of the pipe in pounds-inches per inch of length,
>
> R = radius of the middle line of the pipe wall of the barrel of the pipe in inches,
>
> W = the "ordinary supporting strength" of the sewer pipe calculated in pounds per linear foot of sewer pipe,

F = modulus of rupture of the sewer pipe in pounds per square inch, and
T = thickness of the pipe wall of the barrel of the pipe in inches [3].

The value of 0.20 used in the formula was determined by a large number of tests conducted on curved beams cut from the walls of clay and concrete pipe. "This value of the coefficient agrees well with a theoretical value obtained mathematically by applying the theory of elastic rings to loadings approximating those of the pipe in the tests and in actual ditches" [3].

At this point, there were no standardized tests for testing pipe, but three methods had been developed and used to apply loads to pipe. These included the sand-bearing test, the two-point bearing test, and the three-point bearing test. Any one of these tests could be theoretically used as long as the equation for determining the modulus of rupture was utilized to determine the load.

Although widely used at the time, the sand-bearing test had extremely detailed requirements for testing. The pipe test bedding had to be exactly half the radius of the middle line of the wall at the thinnest point of the pipe wall (pipe wall consistency was not typical given the production methods at the time). Sand passing the No. 4 screen was to be used. The pipe had to be carefully bedded for its full length, above and below, for a quarter of the circumference of its barrel, including any "hubs." The bearing frame could never come in contact with the pipe anytime during the test. The upper surface of the sand in the top bearing had to be struck level with a straight edge, and a strip of cloth was to be used to prevent the loss of sand between the pipe and test frame. The applied load could be placed with either dead weights (in some cases people because they were easier to place and remove) or by applied mechanical means. Given these requirements, it is amazing to think that most of the original pipe-bearing analysis was conducted by this process. It is not surprising that this method was eventually dropped in favor of test methods requiring less preparation time and detail.

The two-edge bearing test method was much easier and less messy than the sand-bearing test method and less difficult to run in a laboratory environment. The two-edge bearing test is essentially what it states. The pipe's halves are marked, and it is placed between two one-inch wide metallic bearings. One can quickly see that the test has one obvious problem—centering a circular, rigid structure between, essentially, two pinching points could result in rolling that, in an interior laboratory, could quickly become an exciting situation. To address this issue, the pipe was placed on a semi-plastic plaster prior to loading. The plaster was then allowed to completely harden to provide some resistance to lateral movement. In larger diameter, higher strength pipe, the loads could become substantial, and a hardened plaster is not going to prevent lateral movement if the pipe is not exactly centered. Although this test method may be the most accurate application of load, its safety and preparation requirements led to the use of a more practical means for evaluating the strength of concrete pipe.

Marston started to compare the sand-bearing test method to the results from the three-point bearing test method in 1911. By 1917, enough correlating testing had been conducted to assess the relative results. It was found that the sand-bearing test method gave direct prediction of the "ordinary supporting strength" of pipe. The three-point bearing test, however, required a conversion factor, "The cracking loads in tests with 'three-point' bearings must be multiplied by 10/7 in computing the 'ordinary supporting strength'" [3]. The more obvious assessment was made in the final recommendation of this report, "Three-point bearings have the advantage of greater rapidity and convenience in the laboratory, and the 'ordinary supporting strength' can be obtained approximately by applying the multiplication factor of 10/7 to their results" [3].

The three-edge bearing test from this point forward became the definitive test by which all concrete pipe was evaluated. These test protocols were ultimately incorporated into ASTM C497, *Standard Test Methods for Concrete Pipe, Manhole Sections, or Tile* [4]. This test method is illustrated in **Fig. 2**.

Three-Edge Bearing Analysis

The performance and consistency of the three-edge bearing test has a long history. When pipe is produced to the minimum requirements in ASTM C76, *Standard Specification for Reinforced Concrete Culvert, Storm Drain, and Sewer Pipe* [5], the pipe will meet or exceed the stated strength for the specified class of pipe. The following test data illustrate this point for standard Class III pipe (1,350 lbf·ft per foot of inside diameter)

FIG. 2 Three-edge bearing machine (U.S. Bureau of Reclamation).

TABLE 1 D-load test report summary for 18-in., 36-in., and 60-in. pipe. ASTM C76, Class III, various plants.

Dia.	ASTM C76 D-load	Actual D-load	Age (Days)	Dia.	ASTM C76 D-load	Actual D-load	Age (Days)
18-in.	1,350D	2,090	6	**36-in.**	1,350D	1,560	13
		3,580	4			1,430	6
		1,820	4			23,240	7
		1,490	5			1,783	1
		2,160	4			2,100	2
		1,610	5			1,807	2
		1,730	2			1,515	13
		2,537	93			1,778	64
		2,691	2			1,900	14
		2,759	5			1,685	9
		2,637	5				
		1,958	2				
		2,799	7				
60-in.	1,350D	1,620	21				
		2,000	2				
		2,210	8				
		1,725	2				
		1,563	11				
		1,405	32				
		1,703	76				
		1,735	21				
		1,723	1				
		1,635	16				

for a standard group of pipe diameters regardless of the production facility. This standardization allows one to evaluate the performance of concrete pipe regardless of the manufacturer or geographical area where the product was produced. It should be noted that, for smaller diameter pipe (i.e., less than 24 in.), the three-edge bearing strengths significantly exceed the targeted D-Load for Class III pipe due to the benefit of compressive thrust strength in high radius, small diameter pipe that is not adequately accounted for in the D-load analysis. This benefit decreases significantly as the pipe diameters increase and is essentially nonexistent in very large diameter pipe, where shear forces exceed any benefit derived from compressive thrust. The three-edge bearing strengths in these cases are very close to those required for the D-load class of pipe.

New Applications of Three-Edge Bearing Design

Although the three-edge bearing test is now more than 100 years old, its utilization has effectively entered the new century mark with further uses and enhancements.

New specifications for fiber reinforcement in concrete pipe have required revisiting the processes associated with conducting the test and new criteria for performance evaluation.

The ASTM C1765, *Standard Specification for Steel Fiber Reinforced Concrete Culvert, Storm Drain, and Sewer Pipe* [6], and ASTM C1818, *Standard Specification for Synthetic Fiber Reinforced Concrete Culvert, Storm Drain, and Sewer Pipe* [7], standards require new procedures for three-edge bearing testing of pipe with fibers. In both new standards, the usual $D_{0.01}$ crack criteria have been replaced by a $D_{service}$. The values, however, for both the $D_{0.01}$ and $D_{service}$ are the same. The use of the same criteria for the previous hundredth-inch and service load criteria essentially makes all the standards interchangeable with the original ASTM C76 standard. The $D_{service}$ requirement, however, is not a 0.01-in. criteria but rather a 1.5 reduction from the ultimate load, which is defined in ASTM C1765 as a 1.5 safety factor.

The two new standards also maintain the existing three-edge bearing ultimate load defined as D_{Test} and D_{ult} in ASTM C1765 and ASTM C1818, respectively. These values, however, have been increased to be uniformly calculated as a 50 % increase over the 0.01-in. or service load values. A comparison of ultimate three-edge bearing loads among these standards would not be a direct correlation to the originally defined ultimate load in ASTM C76, which has some values slightly less than those in the new standards.

An additional new three-edge bearing test has also been included in ASTM C1818. The D_{Reload} test reloads the pipe after it has been tested to ultimately ensure that the pipe does not collapse if it is overstressed because pipe under this standard uses synthetic fibers that can pull out or catastrophically rupture.

This new generation of pipe standards has moved away from the formula-based designs associated with ASTM C76, where required concrete compressive strengths, steel areas, and reinforcement type and positioning were specified. ASTM C1765 and ASTM C1818 are more performance-based standards that essentially allow for unlimited variation in design. So, ironically, rather than the century-old three-edge bearing test being retired, its use is now more critical than ever for the accurate evaluation and design of concrete pipe.

Summary

Without the three-edge bearing test, it would have impossible to develop reinforced precast concrete pipe for sewer applications. This test is really the only means for assessing the final strength of pipe and confirming that it is manufactured as required. It is also one of the only test methods for evaluating a finished product for major infrastructure construction. Most other construction materials have their raw materials certified or their designs analytically evaluated. The three-edge bearing test ensures a specific strength for the finished product that can be validated.

The test's continued use into the twenty-first century indicates it will continue to play a critical role for the next 100 years and beyond.

References

[1] Marston, A. and Anderson, A. O., *The Tests of the Strength of Clay and Cement Drain Tile*, Iowa State College of Agriculture and Mechanic Arts, Ames, IA, 1908.

[2] Marston, A. and Anderson, A. O., "The Theory of Loads on Pipes in Ditches and Tests of Cement and Clay Drain Tile and Sewer Pipe," *Iowa State College of Agriculture and Mechanic Arts, Bulletin No. 31*, Vol. 11, No. 5, February 1913.

[3] Marston, A., Schlick, W. J., and Clemmer, H. F., *The Supporting Strength of Sewer Pipe in Ditches and Methods of Testing Sewer Pipe in Laboratories to Determine Their Ordinary Supporting Strength*, Iowa State College of Agriculture and Mechanical Arts, Ames, IA, 1917.

[4] ASTM C497-16a, *Standard Test Methods for Concrete Pipe, Manhole Sections, or Tile*, ASTM International, West Conshohocken, PA, 2016, www.astm.org

[5] ASTM C76-16, *Standard Specification for Reinforced Concrete Culvert, Storm Drain, and Sewer Pipe*, ASTM International, West Conshohocken, PA, 2016, www.astm.org

[6] ASTM C1765-16, *Standard Specification for Steel Fiber Reinforced Concrete Culvert, Storm Drain, and Sewer Pipe*, ASTM International, West Conshohocken, PA, 2016, www.astm.org

[7] ASTM C1818-15, *Standard Specification for Synthetic Fiber Reinforced Concrete Culvert, Storm Drain, and Sewer Pipe*, ASTM International, West Conshohocken, PA, 2015, www.astm.org

STP 1601, 2017 / available online at www.astm.org / doi: 10.1520/STP160120160114

Josh Beakley[1]

The Evolution of the Application of Highway Live Loads to Buried Concrete Pipe

Citation

Beakley, J., "The Evolution of the Application of Highway Live Loads to Buried Concrete Pipe," *Concrete Pipe and Box Culverts, ASTM STP1601*, J. J. Meyer and J. Beakley, Eds., ASTM International, West Conshohocken, PA, 2017, pp. 28–41, http://dx.doi.org/10.1520/STP160120160114[2]

ABSTRACT

For many years, the application of highway live loads to the surface, and their distribution down to buried concrete pipe, was consistent and reasonably easy to understand. However, near the beginning of the new millennium, the application of highway live loads through soil began an evolution in the United States that resulted in a myriad of changes. The effect of live load on a buried pipe is a result of the application of the live load at the surface, the assumed distribution of that live load through the soil, and furthermore, the dissipation of the load through the structure itself. Modifications have been made to all of these parameters in the American Association of State Highway and Transportation Officials Load and Resistance Factor Design (AASHTO LRFD) *Bridge Design Specifications* within the last two decades. The intent of this paper is to review the history of highway live load design for concrete pipe and to discuss their development within the *AASHTO LRFD Bridge Design Specifications*. Despite the fact that highway live loads themselves have barely changed over the decades, the live load distribution factor has undergone more than one change in the AASHTO codes over the last several years. Meanwhile, the dissipation of the load through the pipe itself was never really addressed within the AASHTO until recently, but a method developed by the concrete pipe industry has been used for years. This has led to inconsistencies with regard to how engineers would address these issues, depending upon which references were consulted for

---

Manuscript received August 24, 2016; accepted for publication November 21, 2016.
[1]American Concrete Pipe Association, 8445 Freeport Pkwy., Irving, TX 75063
[2]ASTM Symposium on *Concrete Pipe and Box Culverts* on December 7, 2016 in Orlando, Florida.

their designs. This paper reviews the history of highway live load design on buried concrete pipe, including tabular and graphical examples of results from the various methods, and provides suggested applications.

Keywords
live loads, buried pipe, concrete pipe, live load distribution, D-Load

Introduction

For many decades, the application of highway live loads to concrete pipe buried below the surface was consistent and reasonably easy to understand. However, at the beginning of the new millennium, the application of highway live loads through soil in the United States began an evolution that resulted in a myriad of changes. Seldom did these changes make the design of concrete pipe any easier or more accurate to perform. The intent of this document is to review the history of highway live load design for concrete pipe and to discuss the evolution of the American Association of State Highway and Transportation Officials Load and Resistance Factor Design (AASHTO LRFD) *Bridge Design Specifications* and the effect the formation of its requirements have had in confusing the issue. Additionally, the potential for improvements through future modifications to the *AASHTO LRFD Bridge Design Specifications* will be discussed.

D-Load Design

The majority of concrete pipe is designed and specified using the D-load concept. This indicates the minimum allowable load in pounds per linear foot of pipe, per linear foot of diameter (lb/ft/ft) to produce a 0.01-in. crack in the pipe when it is tested in the three-edge bearing test apparatus. The three-edge bearing test gets its name because of the three lines of force/resistance it applies to the pipe. Essentially, there is a concentrated load applied at the top of the pipe along its length and two bearing strips underneath the pipe that serve as reaction points. The two bearing strips are spaced 1 in. apart for every foot of internal diameter—just enough to keep the pipe from rolling off the test apparatus. With a much higher concentration of reaction, and no lateral support, the three-edge bearing test is more severe than the load applications in the field. To correlate field loads back to a three-edge bearing test load that produces the same bending stress in the pipe, the field loads—be they earth loads or live loads—are divided by a "bedding factor."

Thus, there is a direct relationship between the field loads and three-edge bearing loads. The higher the load anticipated on the pipe in the field (for the same installation conditions), the higher the D-load requirement for testing the pipe at the plant. To simplify inventories, ASTM C76, *Standard Specification for Reinforced Concrete Culvert, Storm Drain, and Sewer Pipe*, establishes five standard classes of pipe [1]. The classes have been reproduced in **Table 1**.

TABLE 1 ASTM C76 pipe classes and their associated D-load requirements.

Pipe Class	D-load to Produce a 0.01 in. Crack (lb/ft/ft)	D-load to Produce the Ultimate Load (lb/ft/ft)
I	800	1,200
II	1,000	1,500
III	1,350	2,000
IV	2,000	3,000
V	3,000	3,750

The Industry's Method for Many Years

For decades, the American Concrete Pipe Association (ACPA) assumed that highway live loads distributed through the soil at a rate of 1.75 times height (H), where H represents the height of soil cover above the top of the pipe [2]. In other words, for every foot of soil between the surface and the top of the pipe, the horizontal dimension increased by 1.75 ft (see **Fig. 1** and **Table 2**). AASHTO utilized this same distribution in the *AASHTO Standard Specifications for Highway Bridges*, which was the main reference used for highway load design in the United States until the late 1990s [3].

In addition to spreading the load through the soil, the concrete pipe industry also assumed that, upon reaching the pipe, the load distributed through the pipe at the same rate of 1.75 times the vertical distance, where the vertical distance in the pipe is calculated as 0.75 times the outside vertical dimension—the rise (R_o) of the pipe (**Fig. 2**).

FIG. 1 Wheel load spread through soil.

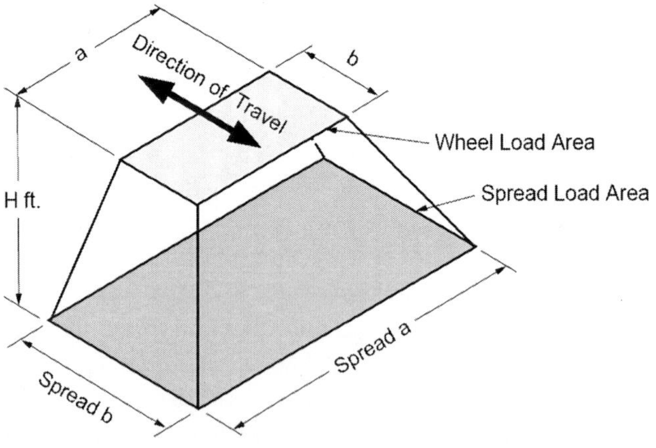

TABLE 2 Specifications and their spread dimensions.

Specification	Soil Type	Wheel Dimensions (in.)		Spread through Soil (ft)	
		a	b	Spread a	Spread b
AASHTO Standard Specifications	N/A	1 or 20 [4]	1 or 10 [4]	a/12 + 1.75H	b/12 + 1.75H
AASHTO LRFD Bridge Design Specifications, 2nd ed. (1998)	Select Granular Backfill	20	10	a/12 + 1.15H	b/12 + 1.15H
	All Other Cases	20	10	a/12 + 1.0H	b/12 + 1.0H
AASHTO LRFD Bridge Design Specifications, 7th ed. (2014)	N/A	20	10	From a/12 + 1.15H for 24-in. ID and below to a/12 + 1.75H for 96-in. ID and above	From b/12 + 1.15H for 24-in. ID and below to b/12 + 1.75H for 96-in. ID and above

Notes: 1. The specification utilizes a "concentrated load," as stated in Article 6.4.1 of the *AASHTO Standard Specifications for Highway Bridges*. Some states utilize a point load (1 in. by 1 in.) while others accept the definition for tire contact area (20 in. by 10 in.) as found in Article 3.30 of the specification as an appropriate "concentrated load."
2. H = earth fill height above the top of the pipe in feet; ID = inner diameter.

FIG. 2 Live load spread through pipe according to ACPA.

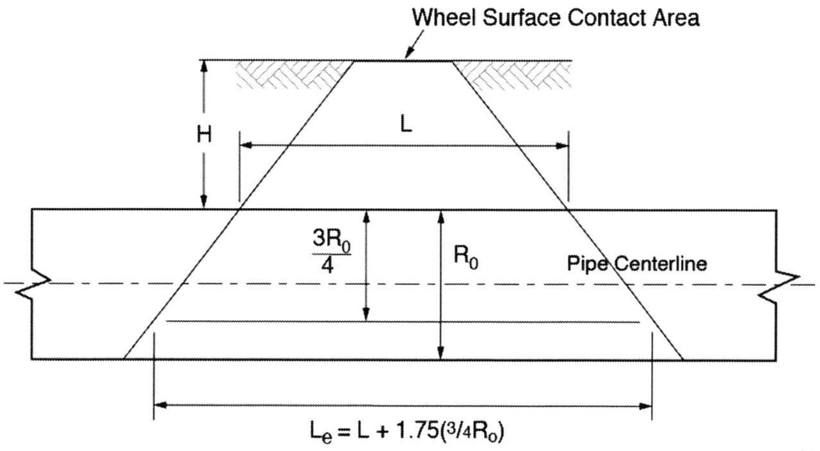

Pipe designed in accordance with this method performed well in the field. However, when researching the history of this design method, there was very little technical substantiation behind the equations. Additionally, for a typical HS20 live load, the resulting D-load requirement for a pipe under shallow fill was much less than the Class III minimum D-load (1,350 lb/ft/ft) required by many states. Because most states require a Class III pipe as a minimum under highways—if not a Class IV, these designs were never really put to a true test of their accuracy. **Table 3** presents the D-load requirements for pipe designed in accordance with the standard practice prior to 1996.

Modifications to the AASHTO Standard Specifications for Highway Bridges Based on Industry Research

In 1996, the AASHTO adopted the standard installations for the embedment and design of concrete pipe. This modification was predicated on improving the modeling of the soil embedment around the pipe and the accuracy of the soil load and soil support provided by it. The standard installations established four basic types of concrete pipe installations (Types 1 through 4) with specific soil categories and compaction levels. Thus, the concrete pipe design community moved away from the vaguer A through D beddings that did not utilize the latest soil and compaction standards toward installations that could be defined by the engineer and understood by the contractor.

The standard installations did not make any modifications to how the live load was distributed through the soil or onto the pipe. However, they did incorporate a new live load bedding factor to account for highly concentrated loads at the top of a pipe. When the pipe is shallow, and the live load pressure is intensely applied

TABLE 3 D-loads based on the Marston Spangler design using a B bedding [4].

	Required D-Load (lb/ft/ft)					
	Pipe Size (in.)					
Fill Height (ft)	12	24	36	48	60	72
1	938	726	478	387	374	367
2	623	483	430	384	359	361
3	578	451	405	380	379	373
4	612	505	451	427	414	413
5	684	600	527	496	480	471

Note: 1. In the earliest standardized design method for concrete pipe, four bedding classifications (A, B, C, and D) were utilized in a design process developed by Marston and Spangler at Iowa State University in the 1930s. The B bedding is considered representative of the typical department of transportation installation.

FIG. 3 Effect of a highly concentrated live load versus a more uniform earth load.

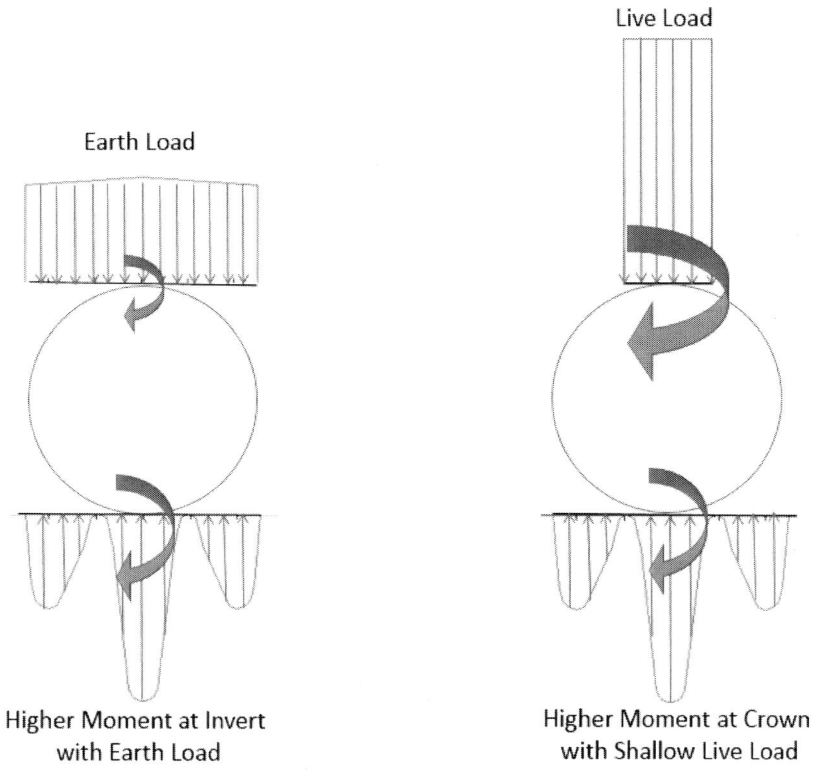

over a small portion of the top of the pipe, the moment at the crown may govern over the moment at the invert (see **Fig. 3**). Interestingly enough, in his research, Spangler suggested the use of a lower bedding factor of 1.5 for the application of concentrated surface loads [5]. For some reason, this recommendation was never incorporated into actual design practice. Thus, although it was known (or at least suspected) for many decades that a live load should have a lower bedding factor, it was not until the late 1990s that it was incorporated into a national standard.

The utilization of lower live load bedding factors resulted in higher D-load requirements. **Table 4** shows the D-load requirements for concrete pipe using a Type 2 installation (reasonably similar to the older B bedding) when the Standard Installations were first incorporated into the *AASHTO Standard Specifications for Highway Bridges* in 1996. When comparing **Table 3** with **Table 4**, one can see how the required D-loads at shallow fill, where live load is the predominant load regardless of the installation type, were increased as a result of the more accurate bedding factor for live load. With the incorporation of the standard installations, increased benefit was given for the earth load bedding factor in addition to the changes made

TABLE 4 D-loads based on the standard installations, using a Type 2 installation.

	Required D-Load (lb/ft/ft)					
	Pipe Size (in)					
Fill Height (ft)	12	24	36	48	60	72
1	1,161	930	761	670	671	684
2	667	585	557	533	530	582
3	529	492	490	495	506	509
4	519	499	506	518	535	555
5	554	539	551	565	585	608

to the live load bedding factors. Thus, as the depth increased, the benefits from the increased earth load bedding factors began to offset the effects of the reduced live load bedding factors. Because it takes longer for the live load to attenuate through the soil to a point beyond the width of the pipe for larger diameter pipes, it takes longer for the live load effect to reduce in larger diameter pipes than it does in smaller diameter pipes. Hence, for a 12-in. diameter pipe, the revised D-loads become less than the initial Spangler D-loads at 3 ft. However, they do not reduce below the Spangler D-loads until depths greater than 5 ft for 36-in. pipe and larger.

Introduction of the *AASHTO LRFD Bridge Design Specifications*

Toward the end of the last century, the *AASHTO LRFD Bridge Design Specifications* [6] became the primary bridge design standard in the United States, and the *AASHTO Standard Specifications for Bridge Design* were no longer maintained. In the development of the *AASHTO LRFD Bridge Design Specifications*, it would seem that little consideration was given to previous live load design practice for buried concrete pipe because the LRFD standard did not continue the practice of distributing the load by a factor of 1.75 times H. Instead, the LRFD bridge design standard incorporated a much smaller distribution of either 1.15 times H for granular soil or 1.0 times H for nongranular soils. This method was based on a 30° distribution through the soil ($\tan(30) = 0.577$, and $2 \times 0.577 = 1.15$). This distribution can be found in many soil textbooks and was assumed appropriate for live load distribution down to the pipe absent any revelation of a technical basis for the 1.75 distribution in use at the time. The 1.0 distribution value was assumed more appropriate for soils with lower internal angles of friction. The result of a smaller live load distribution through the soil is a larger live load pressure at the top of the pipe.

In addition to the live load distribution being reduced (thereby increasing the live load pressure on the pipe), the initiation of the LRFD bridge design specifications also increased the multiple presence factor used in live load design. The multiple presence factor is a factor applied to the service live load to account for the reduction in the probability that multiple lanes will be overloaded at the same time versus just one lane being overloaded. In the AASHTO standard specifications, the multiple presence factor for one or two lanes was 1.0, reducing from this when designing for three or four lanes. After applying the multiple presence factor, the resulting service load would then be multiplied by a load factor of 2.17 for ultimate load design. With the new AASHTO LRFD specifications, the multiple presence factor is 1.2 for one lane and 1.0 for two lanes, continuing to reduce for designs of three or four lanes. After applying the multiple presence factor, the resulting service load would then be multiplied by a load factor of 1.75. When the LRFD bridge design specifications were developed, the explanation used to pacify those folks concerned with the now higher multiple presence factor was that a multiple presence factor of 1.2 paired with a load factor of 1.75 resulted in an ultimate load that was 2.1 times the applied live load. Thus, this was actually a slight reduction in ultimate load in comparison to the 2.17 load factor used in the standard specifications. However, reinforced concrete pipe is most typically specified using the 0.01-in. crack D-load, which is a service load criteria. Because the multiple presence factor is applied to service loads as well as to ultimate loads, the live load used in the typical service load, 0.01-in. design process for concrete pipe is now 20 % higher for a single-lane application when using the *AASHTO LRFD Bridge Design Specifications.*

National Research to Evaluate Live Loads on Buried Structures

The *AASHTO LRFD Bridge Design Specifications* resulted in higher pipe strengths being required even though there had never been any indication that the previous design methods were insufficient (see **Table 5** for the LRFD (1998) D-load values). With the higher D-loads, there was considerable concern that the public was now unnecessarily spending more money to install concrete pipe under roads. Because of these concerns, AASHTO supported a National Cooperative Highway Research Project (NCHRP 15-29) to evaluate live loads on buried structures. This research culminated in NCHRP Report 647, "Recommended Design Specifications for Live Load Distribution to Buried Structures" [7]. This research was supposed to finally solve the issue of how to spread live load through soil and apply it to the pipe appropriately. The report made separate recommendations for the live load application to plastic pipe, corrugated metal pipe, and concrete pipe. This paper only focuses on the application of live load to concrete pipe.

NCHRP Report 647 recommended that the minimum ratio of horizontal live load spread to vertical distance below the surface be 1.15. The research also concluded that this spread ratio was regardless of the soil, and it thus recommended the removal of the 1.0 ratio for nongranular soils that existed in the *AASHTO LRFD Bridge Design Specifications* at that time. The 1.15 ratio was a minimum value for 24-in. pipe and under, while pipe 96 in. in diameter or greater could use a ratio

TABLE 5 D-loads resulting from past, existing, and proposed AASHTO designs.

Type 2 Installation

Depth (ft)	Code	Pipe Inside Diameter (in.)					
		12	24	36	48	60	72
1	LRFD (1998)	1,700	1,050	850	825	800	775
	LRFD (2014)	1,492	1,202	1,244	966	948	850
	Proposed	1,659	1,101	920	944	998	1,003
2	LRFD (1998)	1,050	875	700	650	675	725
	LRFD (2014)	1,322	1,203	1,137	935	875	837
	Proposed	993	906	749	667	717	809
3	LRFD (1998)	800	725	675	625	600	600
	LRFD (2014)	880	818	789	732	696	679
	Proposed	783	732	683	650	618	632
4	LRFD (1998)	725	650	650	650	625	650
	LRFD (2014)	727	690	687	663	650	643
	Proposed	703	670	642	623	634	644
5	LRFD (1998)	700	650	650	650	675	675
	LRFD (2014)	694	665	665	655	654	658
	Proposed	683	660	645	644	657	675

of 1.75. In between these two sizes, the ratio could be linearly interpolated (see **Table 2**). The net result of this was a live load distribution through the soil that was equal to or greater than what was currently required in the *AASHTO LRFD Bridge Design Specifications* (resulting in live load pressures equal to or less than LRFD). However, the live load distribution through the soil was still equal to or less than what was required in the previous *AASHTO Standard Specifications for Highway Bridges* (resulting in live load pressures equal to or greater than the standard specifications).

The other item that greatly affected the live load application as it is applied to the pipe is the distribution factor for the live load through the pipe itself. It should be noted that the 1.75 by 0.75 by pipe rise that was used as a distribution through the pipe by the industry has never been mentioned in the AASHTO design standards. The AASHTO design standards did not cover the distribution of the live load through the pipe, only through the soil. However, it is a logical assumption that a pipe with greater stiffness than the surrounding soil would allow the live load to continue to attenuate through it. For years, the ACPA application of 1.75 by 0.75 by outside vertical dimension of the pipe worked well and was accepted by the engineering community. However, the equations resulting from the NCHRP 15-29 research incorporated a distribution of 0.06 times the inside diameter of the pipe in addition to the distribution through the soil.

The 0.06 times D factor has a significant effect on the design live loads for concrete pipe. There is a tremendous difference between a load spread 0.06 times the rise versus 1.75 × 0.75 = 1.3125 times the rise. Using the simple example of a 12-in. B-wall (2-in. wall) pipe, the ACPA method would spread the load an additional distance of 1.3125 × (12 + 2(2)) = 21 in. versus 0.06 × 12 = 0.72 in. with the method developed in the NCHRP research. Please note that the ACPA factor utilizes the outside diameter, whereas the NCHRP factor utilizes the inside diameter. Although the extreme difference between the two spread dimensions would seem to indicate a lack of justification for the 0.06 times D value, what is more incredulous is to accept that the live load would spread less than 1 in. upon reaching the top of the pipe. The NCHRP research was performed using computer models and could not be justified by physical testing, blaming the differences on "uncertainties in the field tests." Thus, the results have very little substantiation. Nonetheless, these are the LLDF values that have been used by AASHTO since 2012.

Changes to the AASHTO LRFD Bridge Design Specifications in 2012

The application of such a minor spread of load through the pipe would lead to required pipe strengths that were extremely high. The AASHTO Subcommittee on Bridges and Structures, in an attempt to reduce this effect, decided to increase the live load bedding factors recommended in NCHRP Report 647 (see **Table 6** for the live load bedding factors over the years). Please note that the values in the rows in **Table 6** that are labeled "LRFD (2014)" are the current values in the AASHTO. The "proposed" values will be discussed later in this text.

As an additional means of trying to reign in the extremely high application of live loads resulting from the NCHRP research, AASHTO incorporated the same live load distribution through the top of pipes under less than 2 ft of fill that currently existed in Section 4.6.2.10 of the standard for box culverts at these depths,

TABLE 6 Past, existing, and proposed AASHTO live load bedding factors.

		Live Load Bedding Factors					
		Pipe Inside Diameter (in.)					
Depth (ft)	Code	12	24	36	48	60	72
1	LRFD (1998)	2.2	2.2	1.7	1.5	1.4	1.3
	LRFD (2014)	3.2	3.2	2.2	2.2	2.2	2.2
	Proposed	2.2	2.2	1.7	1.5	1.4	1.3
2	LRFD (1998)	2.2	2.2	2.2	2.0	1.8	1.5
	LRFD (2014)	2.4	2.4	2.2	2.2	2.2	2.2
	Proposed	2.2	2.2	2.2	2.0	1.8	1.5
3	LRFD (1998)	2.2	2.2	2.2	2.2	2.2	2.2
	LRFD (2014)	2.4	2.4	2.2	2.2	2.2	2.2
	Proposed	2.2	2.2	2.2	2.2	2.2	2.2

whereby the live load is assumed to distribute through the structure at a value as shown in Eq 1:

$$E = 96 + 1.44S \tag{1}$$

where:

E = distribution of the live load through the top slab of the structure (in.), and

S = span of the structure (ft).

In essence, two wrongs—improper load distribution through the pipe and inappropriate live load bedding factors—were incorporated into the standard to make a right (correct) result. Nevertheless, although some of the live loads were reduced at the fill heights below 2 ft, the deeper fill heights were still greatly punished by the recommendations of NCHRP Report 647 (see **Table 5** for the LRFD (2014) D-load values). Please note that, similar to **Table 6**, the values in the rows labeled "LRFD (2014)" are the current values in AASHTO. The "proposed" values will be discussed later in this text.

Industry Research on Load Distribution Through Concrete Pipes

Regardless of its history of use, the accuracy of the 1.75 times 0.75 times rise factor utilized by the ACPA to distribute load through the pipe was still doubted by some engineers. As mentioned previously, most states require a minimum of a Class III pipe under their highways. Past designs using the ACPA distribution typically resulted in pipe classes well under a Class III pipe, and thus, in most cases, the arbitrary requirement of a Class III pipe overruled the design method used by ACPA when it came to choosing the pipe. Therefore, there are probably few occasions where the ACPA design method was actually put to the test in a field application.

With considerable confusion on the subject, and live load effects on pipe continuing to remain at an all-time high, the ACPA decided to take a look at live loads in a more practical fashion. Rather than apply live loads in a computer model and then explain why they do not match previous test results, the concrete pipe producers got together and performed physical testing of real concrete pipe specimens using a concentrated load similar to a tire footprint. This research incorporated a block 6 in. by 10 in. applied at the very edge of the pipe while it sat in the three-edge bearing rack (see **Fig. 4**). Thus, the resulting establishment of a means to calculate the dissipation of load through the pipe is probably a lower estimate because any soil or pavement above the pipe, which was absent in the test condition, would help to transfer load across the joint.

The test results were published in the 2012 ASCE Pipelines Conference Paper, "Physical Evaluation of the Dissipation of a Concentrated Load When Applied to Reinforced Concrete Pipe" [8]. The resulting equation for the coefficient of live load distribution through the pipe is shown in Eq 2.

FIG. 4 Test of concentrated load at the edge of concrete pipe.

$$Coeff = 242 \times (D_o \times 12)^{-1.97} + 0.855 \quad (2)$$

where:

D_o = outside diameter of pipe (in.), and

Coeff = coefficient to be multiplied by the outside pipe diameter to determine the load distribution through the pipe.

Fig. 5 compares results for the live load distribution through the pipe for three methods; the long standing ACPA method of 1.75 times 0.75 times rise, the latest AASHTO requirement of 0.06 times D_i, and the physically verified distribution found in the ASCE paper. There is some similarity between the ACPA long standing method and the test results for the smaller sizes, thus justifying to some extent the older method. The current AASHTO method can barely be seen at the bottom of the graph because the distribution is so small. The graph of the ACPA distribution increases indefinitely with pipe size. This is because the ACPA literature does not provide a maximum value. However, in practice it is common to see the distribution limited to the length of the pipe or 8 ft, whichever is greater. The proposed distribution from the ASCE paper has a limit of 54 in. This is simply because testing has not been performed on larger size pipe. Thus, there may be a need to increase the limit on the proposed live load distribution for larger pipe as more information

FIG. 5 Live load distribution through the pipe based on currently existing criteria.

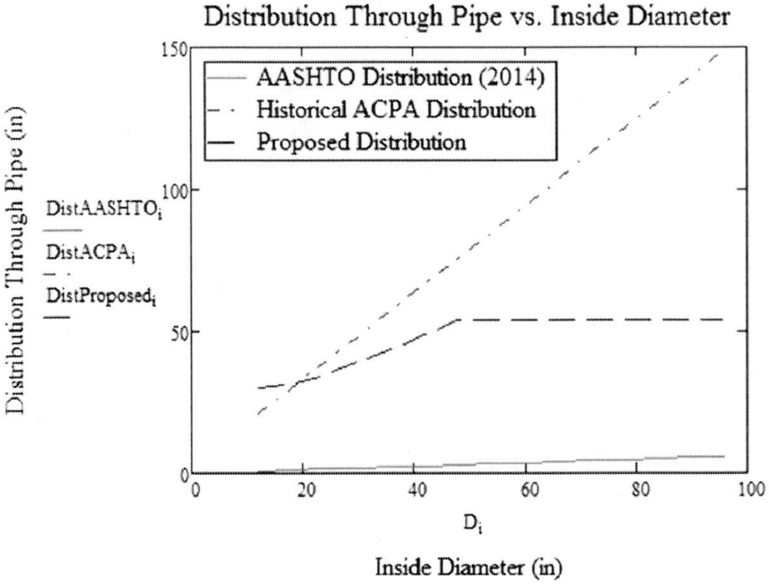

becomes available. Although the ACPA testing was not carried out through NCHRP funding, the resulting live load distribution would seem to make more sense simply on a practical level than any other suggested method of live load through the pipe previously considered. It has physical testing as its basis, and it results in distributions that go beyond mere inches.

The concrete pipe industry has developed proposed changes to the AASHTO specifications that would go back to the old live load bedding factors, which have some technical justification, while incorporating live load distribution through the pipe in accordance with Eq 2 for the distribution coefficient. The live load distribution through the soil itself would remain as it currently exists in the AASHTO specifications with the minimum value of 1.15 and maximum value of 1.75 with linear interpolation for sizes between 24 in. and 96 in.

Summary

The application of live loads through soil down to the pipe, and then subsequently through the pipe, has seen numerous changes in the last 20 years. A lot of these changes have been significant. It is doubtful that the current live load design requirements for buried concrete pipes are the final chapter in this saga. The continuous changes to live load design of buried concrete pipe over the last 20 years have made it difficult for engineers to keep up with the latest *AASHTO LRFD Bridge Design Specifications* and very hard to maintain design software to the latest requirements. Additionally, the situation leads to some hesitancy in updating software because the resulting designs have been overly conservative, and one would hope the LRFD code would be corrected before engineers would start using software to perform designs that will needlessly cost the owners more money. The latest proposals from the concrete pipe industry should help to bring some practicality to live load design of buried concrete pipe and hopefully deliver some stability to a topic that has been unsettled for quite a while.

References

[1] ASTM C76-16, *Standard Specification for Reinforced Concrete Culvert, Storm Drain, and Sewer Pipe*, ASTM International, West Conshohocken, PA, 2016, www.astm.org

[2] *Concrete Pipe Design Manual*, American Concrete Pipe Association, Irving, TX, 1996.

[3] *AASHTO Standard Specifications for Highway Bridges*, 16th ed., American Association of State Highway and Transportation Officials, Washington, DC, 1996.

[4] *AASHTO LRFD Bridge Design Specifications*, 2nd ed., American Association of State Highway and Transportation Officials, Washington, DC, 1998.

[5] Spangler, M. G., *The Supporting Strength of Rigid Pipe Culverts*, Bulletin 112, Iowa Engineering Experiment Station, Ames, IA, 1933.

[6] *AASHTO LRFD Bridge Design Specifications*, 7th ed., American Association of State Highway and Transportation Officials, Washington, DC, 2014.

[7] NCHRP Report 647, *Recommended Design Specifications for Live Load Distribution to Buried Structures*, Transportation Research Board, Washington, DC, 2010.

[8] Beakley, J., "Physical Evaluation of the Dissipation of a Concentrated Load When Applied to Reinforced Concrete Pipe," presented at the *Pipelines 2012: Innovations in Design, Construction, Operations, and Maintenance—Doing More With Less*, Miami Beach, FL, August 19–22, 2012, American Concrete Pipe Association, Irving, TX, 2012, http://dx.doi.org/10.1061/9780784412480.126#sthash.iqN2ISTE.dpuf

STP 1601, 2017 / available online at www.astm.org / doi: 10.1520/STP160120160125

Ashley Wilson,[1] Ali Abolmaali,[2] Yeonho Park,[3]
and Emmanuel Attiogbe[4]

Research and Concepts Behind the ASTM C1818 Specification for Synthetic Fiber Concrete Pipes

Citation

Wilson, A., Abolmaali, A., Park, Y., and Attiogbe, E., "Research and Concepts Behind the ASTM C1818 Specification for Synthetic Fiber Concrete Pipes," *Concrete Pipe and Box Culverts, ASTM STP1601*, J. J. Meyer and J. Beakley, Eds., ASTM International, West Conshohocken, PA, 2017, pp. 42–49, http://dx.doi.org/10.1520/STP160120160125[5]

ABSTRACT

ASTM C1818, *Standard Specification for Synthetic Fiber Reinforced Concrete Culvert, Storm Drain, and Sewer Pipe*, has recently been approved as a performance-based ASTM specification based on multiple tests conducted by the authors on synthetic fiber reinforced concrete pipes (SYN-FRCP). The initial testing included the production and three-edge bearing testing of SYN-FRCP pipes, ranging in diameter from 18 to 36 in. (450 to 900 mm) with varying fiber dosages. More than 250 tests were conducted, and fiber dosages for different diameters and wall thicknesses were identified for achieving certain pipe class strengths. Material testing, including cylinders (ASTM C39, *Standard Test Method for Compressive Strength of Cylindrical Concrete Specimens*) and beams (ASTM C1609, *Standard Test Method for Flexural Performance of Fiber-Reinforced Concrete [Using Beam with Third-Point Loading]*), was also performed at different fiber dosages. Due to the time-dependent behavior of synthetic fibers, long-term field and laboratory tests were conducted for 10,000 hours' duration

Manuscript received September 7, 2016; accepted for publication January 17, 2017.
[1]Forterra Building Products, 9405 Alton Ct., Henderson, CO 80640 http://orcid.org/0000-0002-7104-0822
[2]University of Texas at Arlington, Dept. of Civil Engineering, PO Box 19308, Arlington, TX 76010
[3]University of Texas at Arlington, Dept. of Civil Engineering, PO Box 19308, Arlington, TX 76010 http://orcid.org/0000-0002-2014-7494
[4]BASF Corp., 23700 Chargrin Blvd., Beachwood, OH 44122 http://orcid.org/0000-0001-5525-5623
[5]ASTM Symposium on *Concrete Pipe and Box Culverts* on December 7, 2016 in Orlando, Florida.

to determine the long-term serviceability factor for the synthetic fiber concrete pipes. This long-term property was introduced as a part of ASTM C1818 and is required to be determined by fiber producers through long-term laboratory tests using their respective fibers in concrete pipes. This paper will present the concept behind the ultimate load, which is the load at the vicinity of the first crack. This paper will also introduce reasons behind the tests and design concepts described in the specification.

Keywords

concrete pipe, synthetic fiber, synthetic fiber reinforced concrete pipe, pipe, long-term laboratory testing

Introduction

It is well-known that the presence of fibers in concrete enhances the material's performance, including post-crack control, impact resistance, and a higher energy absorption (defined as a function of the area under the load versus deflection curve) after the initial crack occurs. According to American Concrete Institute (ACI) 544.1R-96 (2002) [1] and the American Association of State Highway and Transportation Officials-Associated General Contractors of America-American Road and Transportation Builders (AASHTO-AGC-ARTBA) Joint Committee (2001) [2], ultimate flexural strength was found to increase when the volume fraction of steel fibers increased within the practical range. An increase in the flexural strength may even be possible at higher volume fractions. For the first time in the United States, concrete pipes reinforced with synthetic (polypropylene) fibers have been introduced by Wilson and Abolmaali [3].

This study, along with the long-term study completed by Park, Abolmaali, and Attiogbe [4], was used in the development of ASTM C1818-15, *Standard Specification for Synthetic Fiber Reinforced Concrete Culvert, Storm Drain, and Sewer Pipe* [5]. The theories leading into this specification are discussed in detail.

Previous steel fiber research was conducted by Abolmaali et al. [6] and, as a result, ASTM C1765, *Standard Specification for Steel Fiber Reinforced Concrete Culvert, Storm Drain, and Sewer Pipe* [7], was created as the steel fiber reinforced concrete pipe (SFRCP) specification. The second phase of this research began and focused primarily on synthetic fiber reinforced concrete pipe (SYN-FRCP) due to the noncorrosive properties of synthetic fibers. Synthetic fibers have been known to excel in slab-on-grade, wall systems, and shotcrete applications. Previous investigations by Song, Hwang, and Sheu [8] and Alhozaimy, Soroushian, and Mirza [9] have shown that polypropylene synthetic fibers have increased compressive strengths, toughness, splitting tensile strength, and impact resistance when applied in concrete applications. Due to past research and the success of the steel fibers in concrete pipe applications, research was completed for SYN-FRCP.

Pipe Design (ASTM C1818, Chapter 8)

ASTM C1818-15 [5] is a performance-based specification that rests on the design philosophy described in the *Concrete Pipe Handbook* [10]. The testing procedures adopted for traditional reinforced concrete pipe (RCP) in ASTM C497, *Standard Test Methods for Concrete Pipe, Manhole Sections, or Tile* [11], and although part of the testing procedures are the same for SYN-FRCP, the design philosophy is very different. The new SYN-FRCP specification [5] is still considered to be classified as a rigid pipe system, similar to the ASTM C1765 [7] and ASTM C497 [11] specifications. The pipe is placed on two longitudinal parallel strips of hard rubber or wood fastened directly to a rigid base, and the load is applied through an upper bearing strip of a rigid wood beam with an attached hard rubber strip along the pipe length. Evaluation and classification (Class I to Class V) of the strength of the precast concrete pipe can be determined based on the three-edge bearing test results. Traditional RCP is D-load tested until the first visible crack is observed, which is defined as a 0.01-in. (0.254 mm) crack, 1 ft. (0.3 m) in length traveling along the length of the pipe. This crack is commonly first observed in the crown of the pipe and then is quickly followed by the invert and spring line locations of the pipe. The D-load is commonly referred to as the $D_{0.01}$ and is a 1.5 factor of safety of $D_{ultimate}$. Both the $D_{0.01}$ and $D_{ultimate}$ load requirements for pipe Classes I, II, III, IV, and V are defined in ASTM C76, *Standard Specification for Reinforced Culvert, Storm Drain, and Sewer Pipe* [12], and are summarized in **Table 1**.

For the new specification, ASTM C1818 [5], $D_{ultimate}$ is the ultimate load beyond what the pipe is able to resist, meaning the pipe has been loaded enough that it cracks. However, after the $D_{ultimate}$ is reached, the pipe is still able to deflect without collapse due to the synthetic fiber reinforcement. The service load is defined as $D_{service}$ and is a 1.5 factor of safety applied to $D_{ultimate}$. Therefore, $D_{service}$ is equal to two-thirds of $D_{ultimate}$. Due to cracking first occurring at the ultimate loading, SYN-FRCP will not crack in service applications.

SYN-FRCP testing will be performed on a standard D-load testing apparatus. Once the $D_{ultimate}$ load has been reached, the load will be removed from the pipe

TABLE 1 Pipe class definitions based on ASTM C76 and ASTM C1818.

	RCP = ACSM C76		SYN-FRCP - ASTM C1818	
Pipe Class	$D_{0.01}$ - Load lb/ft/ft of Diameter (kN/m/m of Diameter)	$D_{ultimate}$ - Load lb/ft/ft of Diameter (kN/m/m of Diameter)	$D_{service}$ - Load lb/ft/ft of Diameter (kN/m/m of Diameter)	$D_{ultimate}$ - Load lb/ft/ft of Diameter (kN/m/m of Diameter)
I	800 (38)	1200 (57)	800 (38)	1200 (57)
II	1000 (48)	1500 (72)	1000 (48)	1500 (72)
III	1350 (65)	2000 (96)	1350 (65)	2025 (97)
IV	2000 (96)	3000 (144)	2000 (96)	3000 (144)
V	3000 (144)	3750 (180)	3000 (144)	4500 (216)

and then reloaded up to a minimum of $D_{service}$ or D_{reload}. In order for the SYN-FRCP to be accepted, the D_{reload} values must be greater than or equal to the $D_{service}$ value. The D_{reload} load will then be held constant for 1 min to ensure the pull-out characteristics of the synthetic fibers are sufficient. D_{reload} is described in more detail in a subsequent section. Based on the performance of the SYN-FRCP in the D-load testing, a representation of the conceptual design philosophy compared to the traditional RCP is shown in **Fig. 1**. The key difference is that SYN-FRCP is uncracked under service loading whereas RCP is cracked. The D-load testing is considered the performance-based proof required for SYN-FRCP to meet ASTM C1818 [5]. Similar to ASTM C76 [12], ASTM C1818 [5] also has prescribed values required for $D_{service}$ and $D_{ultimate}$, which are described in **Table 1**.

Synthetic Fiber-Concrete Matrix Qualification Testing (ASTM C1818, Chapter 9)

Due to the time-dependent behavior of synthetic fibers, there is a requirement in ASTM C1818 [5] in which the long-term behavior of the fibers must be determined. The long-term serviceability factor is α, which is the value assumed to be extrapolated to the 100-year strength of the fiber-concrete matrix. Long-term testing was completed in order to verify the deflection observed in cracked pipes under constant loading. Two different types of tests were completed: field testing described in Park, Abolmaali, and Attiogbe [4] and lab testing described in Attiogbe, Abolmaali, and Park [14]. The field testing under sustained loading was performed to account

FIG. 1 (a) Three-edge bearing test setup and (b) conceptual design philosophy of RCP and FRCP [13].

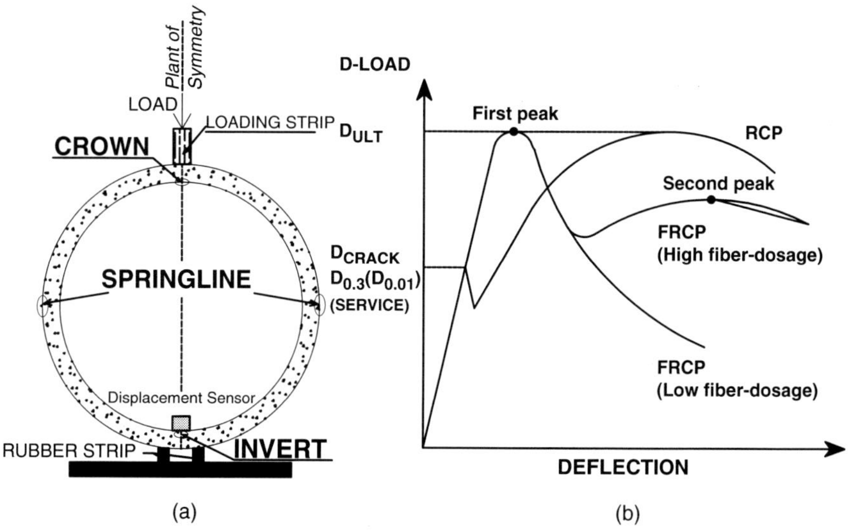

FIG. 2 Long-term testing simulation [14].

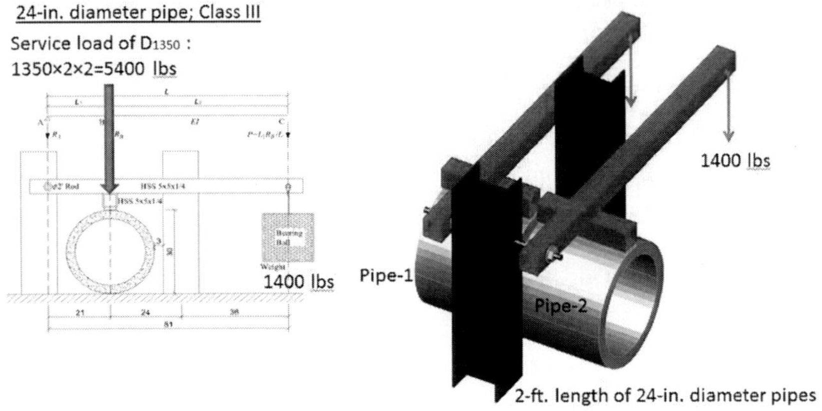

for the effect of soil pressure [4]. In the lab testing, an apparatus was set up to first load the pipe to ultimate, where the crack was first observed; then the pipe was immediately unloaded. Following the initial crack, the pipe was then reloaded to the appropriate service load and a constant load was applied for 10,000 h. During this time, the deflection was monitored. **Fig. 2** shows an example of the lab testing for a 24-in. diameter pipe being tested at a service load equal to 1,350 lb.

The loading protocol for the lab testing is shown in **Fig. 3**. Pipe specimens were loaded in the three-edge-bearing test until the first visible crack was observed at the crown or invert of the pipe (or both). After the initial pre-cracking of the pipes, the applied loads were immediately removed, and the pipes were placed in the long-term test setup. Then, sustained service load ($D_{Service}$) was applied to each pipe.

FIG. 3 Loading protocol.

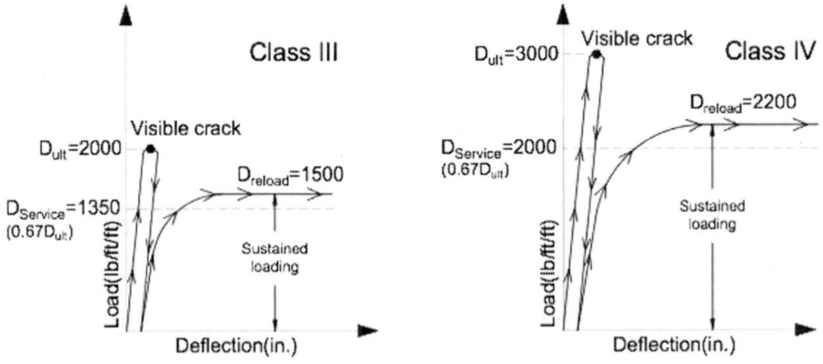

Based on ASTM C1818 [5], FRCP design strength ($D_{service}$) should reach 67 % of the ultimate strength (D_{ult}) of the pipes. Conservatively, in the lab testing, a load level higher than $D_{service}$ (D_{reload} as shown in **Fig. 3**) was applied to the pipes in the long-term testing.

From the lab testing, it was concluded that, after the initial 500 h to 1,000 h of loading, the deflection observed on the pipe was leveled off to the point where no significant increases were noticed for the remainder of the test. Also, the deflection was within 0.7 % difference of the inside diameter when comparing uncracked and cracked pipe tests.

Per ASTM C1818, Section 9.2 [5], the average value of α must be 0.9 or higher, and in no case should any one test be less than 0.8, with α being equal to the final extrapolated inside vertical dimension of the pipe (ID_f) divided by the initial inside vertical dimension of the pipe (ID_o). Per ASTM C1818 [5], measurements of the vertical dimension of the pipe were recorded at the increments shown in **Table 2**. Recording of measurements could cease any time after 100 h, provided that the difference between the last measurement and the one preceding it was less than 0.5 %. However, the load should remain on the pipe for at least 10,000 h to test against brittle failure. At no point during the testing should any crack on the interior or exterior of the pipe wall exceed 0.125 in. for a length of 1 ft or greater. Crack widths greater than 0.125 in. were deemed a failure of the pipe in this test. Provided the pipe did not fail within 10,000 h, the long-term serviceability factor may be established on the basis of the ratio of the final (ID_f) and initial (ID_o) inside vertical dimensions of the pipe.

The purpose of the α factor is to account for accidental overloads in service conditions and to ensure the pull-out capacity of the synthetic fibers. A minimum of three test specimens must be tested and must pass this testing without failure to meet the specification. The tests by Park, Abolmaali, and Attiogbe [4] and Attiogbe, Abolmaali, and Park [14] verified that macro SYN-FRCP can perform well in service if the fiber provides sufficient long-term post-crack strength as evidenced by a long-term serviceability factor α value of ≥ 0.9. Factor $\langle$ is a direct measure of creep deformation in the concrete-fiber composite [14].

TABLE 2 Incremental measurement.

Hours	Measurements Taken at Least
0 to 20	Every 1 h
20 to 40	Every 2 h
40 to 60	Every 4 h
60 to 100	Every 8 h
100 to 600	Every 24 ± 6 h
600 to 6,000	Every 48 ± 10 h
After 6,000	Every week

Pipe Proof of Design Testing (ASTM C1818, Chapter 10)

In order to establish the proof of design, a series of tests must be completed on sample pipes. All testing for proof design should be completed in accordance with ASTM C497, *Standard Test Methods for Concrete Pipe, Manhole Sections, or Tile* [11], and three to seven sample pipes will be tested to represent the typical group of pipes. Once the proof D-load test has been completed, the load will be removed and reloaded to D_{reload}, as described previously. ASTM C1818 [5] describes D_{reload} as equal to $D_{service}/\alpha$. If the three to seven tests verify that each pipe is able to pass the D_{reload}, then the standard deviation(s) of the tests will be calculated to establish the average of $D_{ultimate}$, per ASTM C1818, Section 10.4.2 [5]. If the arithmetic mean is equal to or greater than the minimum allowable arithmetic mean, then the pipes will pass the requirement. Any variation in the material or manufacturing will result in the testing being redone so that the results represent the actual typical pipe.

Conclusions

A performance-based testing procedure has been developed for SYN-FRCP for the first time in the United States. The goal of this technical paper was to compare the design philosophy for ASTM C1818 [5] to that of the traditional reinforced concrete pipe standard, ASTM C76 [12]. A key change is that SYN-FRCP is designed to not crack while in service. Long-term testing must be completed to ensure the fiber-concrete matrix has adequate pull-out capacity and is able to withstand loading in the event the pipe does crack due to unforeseen events. The scope of this standard is limited to pipe diameters of 12 in. (300 mm) to 48 in. (1,200 mm).

ACKNOWLEDGMENTS

Special thanks to the ASTM C13 technical group for their technical input and assistance.

References

[1] ACI Committee 544, *State-of-the-Art Report on Fiber Reinforced Concrete (ACI 544.1R-96)*, American Concrete Institute, Farmington Hills, MI, 2002.

[2] AASHTO-AGC-ARTBA Joint Committee, *The Use and State-of-the-Art Practice of Fiber Reinforced Concrete (Task Force 36)*, American Association of State Highway and Transportation Officials-Associated General Contractors of America-American Road and Transportation Builders, AASHTO, Washington, DC, 2001.

[3] Wilson, A. and Abolmaali, A., "Performance of Synthetic Fiber-Reinforced Concrete Pipes," *J. Pipeline Syst. Eng. Pract.*, Vol. 5, No. 3, 2014, 04014002, http://dx.doi.org/10.1061/(ASCE)PS.1949-1204.0000166

[4] Park, Y., Abolmaali, A., and Attiogbe, E., "Time Dependent Behavior of Synthetic Fiber Reinforced Concrete Pipes Under Long-Term Sustained Loading," *J. Transp. Res. Board*, Vol. 2407, 2014, pp. 71–79.

[5] ASTM C1818-15, *Standard Specification for Synthetic Fiber Reinforced Culvert, Storm Drain, and Sewer Pipe*, ASTM International, West Conshohocken, PA, 2010, http://dx.doi.org/10.1520/C1818-15

[6] Abolmaali, A., Mikhalova, A., Wilson, A., and Lundy, J., "Performance of Steel Fiber-Reinforced Concrete Pipes," *Transp. Res. Rec.*, Vol. 2313, No. 1, 2012, pp. 168–177, http://dx.doi.org/10.3141/2313-18

[7] ASTM C1765-13, *Standard Specification for Steel Fiber Reinforced Concrete Culvert, Storm Drain, and Sewer Pipe*, ASTM International, West Conshohocken, PA, 2010, http://dx.doi.org/10.1520/C1765

[8] Song, P. S., Hwang, S., and Sheu, B. C., "Strength Properties of Nylon- and Polypropylene-Fiber-Reinforced Concretes," *Cement Concrete Res.*, Vol. 35, No. 8, 2005, pp. 1546–1550.

[9] Alhozaimy, A. M., Soroushian, P., and Mirza, F., "Mechanical Properties of Polypropylene Fiber Reinforced Concrete and the Effects of Pozzolanic Materials," *Cement Concrete Comp.*, Vol. 18, No. 2, 1996, pp. 85–92.

[10] Peckworth, H. F., *Concrete Pipe Handbook*, American Concrete Pipe Association, Chicago, IL, 1958.

[11] ASTM C497-16, *Standard Test Methods for Concrete Pipe, Manhole Sections, or Tile*, ASTM International, West Conshohocken, PA, 2010, http://dx.doi.org/10.1520/C0497-16

[12] ASTM C76-15a, *Standard Specification for Reinforced Culvert, Storm Drain, and Sewer Pipe*, ASTM International, West Conshohocken, PA, 2010, http://dx.doi.org/10.1520/C0076-15A

[13] Park, Y., Abolmaali, A., Mohammadagha, M., and Lee, S., "Structural Performance of Dry-Cast Rubberized Concrete Pipes with Steel and Synthetic Fibers," *Constr. Build. Mater.*, Vol. 77, 2015, pp. 218–226.

[14] Attiogbe, E., Abolmaali, A., and Park, Y., "Polypropylene Fiber-Reinforced Concrete Pipes Under Sustained Loading," presented at the *Ninth RILEM International Symposium on Fiber Reinforced Concrete*, Vancouver, Canada, September 19–21, 2016, BEFIB, Vancouver, Canada.

STP 1601, 2017 / available online at www.astm.org / doi: 10.1520/STP160120160121

Corey L. Haeder[1]

History of Reinforced Concrete Low-Head Pressure Pipe Design

Citation

Haeder, C. L., "History of Reinforced Concrete Low-Head Pressure Pipe Design," *Concrete Pipe and Box Culverts, ASTM STP1601*, J. J. Meyer and J. Beakley, Eds., ASTM International, West Conshohocken, PA, 2017, pp. 50–58, http://dx.doi.org/10.1520/STP160120160121[2]

ABSTRACT

ASTM C361, *Standard Specification for Reinforced Concrete Low-Head Pressure Pipe*, was first published in 1955. At that time and for several decades, the U.S. Bureau of Reclamation (USBR) published a similar document (USBR M-1, *Standard Specifications for Reinforced Concrete Pressure Pipe*) based on a similar set of design processes but with differing loading and design assumptions. Consolidation of the two standards was desired by industry and the USBR. Through improved analysis methods and further evolved rational design procedures that take into account improved material properties, the current ASTM C361 standard incorporates the present state of these design concepts and has been adopted by the USBR. The evolution of the design process is presented for user reference.

Keywords

reinforced, concrete, head, pressure, pipe, design, history

Introduction

Reinforced concrete low-head pressure pipe (RCLHPP) is a product that has been used by the irrigation and drainage industries for more than 50 years. This product is intended for use in pipelines that have internal hydrostatic pressures up to a

---

Manuscript received September 1, 2016; accepted for publication November 3, 2016.

[1]Forterra Building Products, Engineering Dept., 6655 Wedgwood Rd., Maple Grove, MN 55311, http://orcid.org/0000-0002-2080-9734

[2]ASTM Symposium on *Concrete Pipe and Box Culverts* on December 7, 2016 in Orlando, Florida.

125-ft head and buried under less than 20 ft of fill, although special designs may be prepared for higher earth covers. Two similar standards that are domestically available for specification of the design and production aspects of RCLHPP are ASTM C361, *Standard Specification for Reinforced Concrete Low-Head Pressure Pipe* [1], and the U.S. Bureau of Reclamation publication, USBR M-1, *Standard Specifications for Reinforced Concrete Pressure Pipe* [2]. A design manual is available that is published by the American Water Works Association (AWWA) entitled, *Concrete Pressure Pipe—Manual of Water Supply Practices*, which is also known as AWWA M9 [3]. This standard provides guidance for similar products but in a different manner than the ASTM and USBR standards; as such, it is not included in this discussion because it is intended for higher head pressure pipe.

The ASTM and USBR documents were developed under separate but collaborative pathways from their inception but recently have converged to a single standard method. The background and history of each standard will be discussed in this document, which will also serve as a road map to connect the prior versions of each document to the current ASTM C361 standard.

For reference, typical nomenclature to describe the pipe sections specified under each document incorporate the same logic as follows:

1. Fill height is indicated by the letter portion of the designation:
   a. A = 5-ft fill
   b. B = 10-ft fill
   c. C = 15-ft fill
   d. D = 20-ft fill
2. Internal pressure is indicated by the number portion of the designation:
   a. 36A25 = 36-in. diameter, 5-ft fill with 25-ft internal static head pressure.
   b. 72D100 = 72in. diameter, 20-ft fill with 100-ft internal static head pressure.
   c. Pressure classes start at 25 ft and are standardized in the design tables for 25 ft increments to 125 ft max for ASTM C361 and 150 ft max for the USBR M-1 standard.
   d. Hydrostatic head, as listed in the class of the pipe, is measured to the horizontal centerline of the pipe.

The evolution of each standard will be discussed separately followed by a description of the process followed to attain agreement between the ASTM procedures and the USBR design standards in the current version of ASTM C361.

ASTM C361

The first version of ASTM C361 was published in 1955. Two historical versions were available to the author (dated 1970 and 1990). These versions contain the following assumptions and tables:

1. Tabulated steel areas—single set of assumptions.
2. Required concrete compressive strength = 4,500 psi.
3. Required reinforcement yield strength = 40,000 psi.
4. Assumed 90° bedding angle achieved during installation.

5. Bedding per the detail shown in **Fig. 1**.
6. Stress analysis determined using methods developed by Olander [4].
7. Loading combination for ultimate strength design cases
   a. Pipe weight + earth load + internal water load + internal hydrostatic head
8. Unit weight of earth fill = 100 pcf + 20 pcf * (fill height/outer diameter [OD]), 150 pcf max.
9. Load factor = 1.8 (all loads/limit states).
10. Capacity reduction factor = 1.0 (all loads/limit states).
11. Allowable steel stress for hydrostatic head only = 17,000 psi − 35 * hydrostatic head
    a. Head is measured at the centerline of the pipe.
    b. Checking hypothetical bursting case is to prevent cracking entirely through the wall.
12. Elliptical reinforcement = 1.6 times steel area required for hydrostatic head.
13. Concrete tensile stress = (0.433 * hydrostatic head * inner diameter [ID])/(two times wall thickness) $\leq$ 325 psi.

Starting in the early 2000s, the ASTM C13 Subcommittee C13.04 began discussing an update to the design procedures included in ASTM C361, Appendix X2, with the goal of adoption of the ASTM C361 standard by the USBR as a replacement for their M-1 standard. These updates were to include benefits of the following:

1. Improved concrete and steel material strengths
2. Rational design procedures including multiple loading combinations, crack control considerations, and shear/radial tension checks

FIG. 1 Pipe bedding.

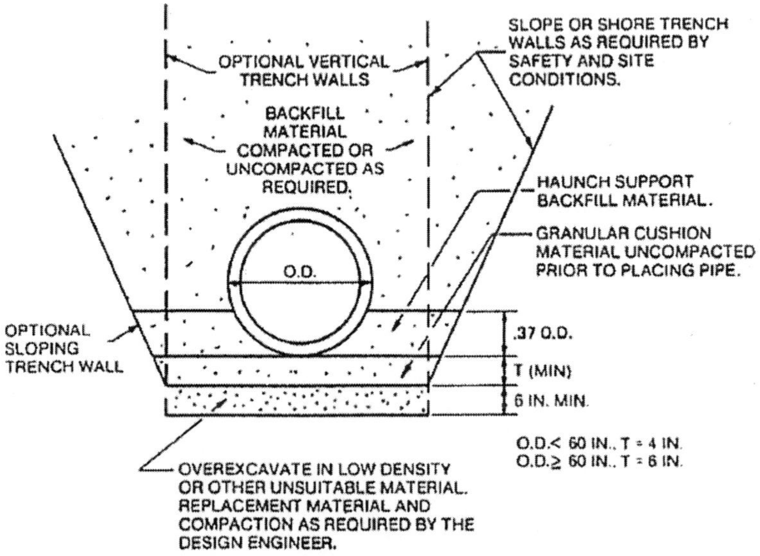

3. Incorporation of additional USBR design preferences for earth load and bedding angle assumptions

As a result of the lengthy discussion and debate, in 2011 an updated version of ASTM C361 was approved for publication. This new version of the standard included the following changes/similarities:

1. Tabulated steel areas based on two sets of assumptions:
   a. ASTM C361, Table 1 = 5,000 psi concrete/40 ksi steel. This table was developed to provide a reasonable similar connection between the new version of the ASTM C361 standard and previous versions recognizing the excellent performance history of this pipe based on earlier design criteria and the fact that some manufacturers use grade 40 reinforcement in some cases. Comparison of selected steel areas is included in Table 1 of this document.
   b. ASTM C361, Table 2 = 5,000 psi concrete/60 ksi steel. This option was desired by industry to account for typical material strengths used in modern manufacturing that may exceed 75 ksi or 80 ksi. A compromise to realize the benefit of increasing from 40 ksi to 60 ksi was the final agreement among the committee members. Selected steel areas for this option are also included in Table 1 of this document.
2. Required concrete compressive strength—two options:
   a. 5,000 psi for ASTM C361, Tables 1 and 2, increased from 4,500 psi to capture typical precast practice.
   b. 6,000 psi for designs that require additional resistance to shear or radial tension.
3. Required reinforcement yield strength—two options:
   a. 40 ksi for ASTM C361, Table 1—in order to provide a calibration between the historical performance of former ASTM C361 design tables based on the assumptions described previously and the expanded rational

TABLE 1 Comparison of tabulated steel areas for selected sizes/classes.

Size	Class	Inner/Outer Steel Area					
		ASTM C361-70	ASTM C361-92	USBR M1-74	USBR M1-91	ASTM C361-11 (T1)	ASTM C361-11 (T2)
12	A25	0.07	0.07	0.07	0.08	0.07	0.05
	D125	0.32	0.32	0.31	0.33	0.31	0.31
36	A25	0.17/0.12	0.17/0.12	0.17/0.12	0.21/0.13	0.18/0.12	0.12/0.08
	D125	0.75/0.51	0.75/0.51	0.75/0.51	0.82/0.54	0.70/0.51	0.54/0.39
60	A25	0.35/0.23	0.35/0.23	0.35/0.23	0.45/0.27	0.37/0.25	0.25/0.17
	D125	1.50/1.00	1.50/1.00	1.50/1.00	1.74/1.13	1.38/0.99	0.92/0.66
84	A25	0.53/0.35	0.53/0.35	0.53/0.35	0.69/0.40	0.57/0.38	0.38/0.26
	D125	2.08/1.40	2.08/1.40	2.08/1.40	2.44/1.56	**1.96/1.40 S**	**1.36/0.93 S**
108	A25	0.71/0.47	0.71/0.47	0.71/0.47	0.93/0.52	0.77/0.51	0.52/0.34
	D125	2.46/1.67	2.46/1.67	2.46/1.67	2.89/1.84	**2.45/1.74 S**	**2.13/1.16 S**

Note: Bold values indicate designs that require shear reinforcement according to ASTM C361.

design methods of the new standard, the steel yield strength starts at 40 ksi. The load factors detailed in the following were then parametrically calibrated to yield reinforcement areas that were reasonably close to the former versions of ASTM C361.

   b. 60 ksi for ASTM C361, Table 2, was incorporated to capture benefit of available materials in modern manufacturing processes using the same calibrated load factors resulting from the development of Table 1 of this document.

4. Bedding angle assumptions:
   a. 90° for earth, water, and live loads.
   b. 45° for pipe weight—lesser angle was conservatively assumed for the determination of pipe dead load stresses given that the method of installation will tend to lock in the dead load stresses before side fill is placed.
5. Bedding detail unchanged.
6. Stress analysis determined using methods developed by Olander [4].
7. Loading combinations for ultimate strength design:
   a. Condition 1 = internal hydrostatic pressure only.
   b. Condition 2 = earth, pipe, and water weight with no hydrostatic pressure.
   c. Condition 3 = combination of external and internal loads concurrently.
8. Unit weight of earth fill = 120 pcf + 24 pcf * (fill height/OD), 168 pcf max—increased per request of the USBR to match with field experience but maximum reduced from 180 pcf to 168 pcf.
9. Load factors:
   a. Flexure, for internal pressure = 1.5, reduced from 1.8.
   b. Flexure, for dead loads = 1.6 if load generates tensile thrust, reduced from 1.8.
   c. Flexure, for dead loads = 1.0 if load generates compressive thrust because it is not conservative to factor a load that is beneficial to the design stresses and would reduce required steel areas.
   d. Shear, for all loads = 1.3 and applies to diagonal tension and radial tension.
10. Capacity reduction factor:
    a. Flexure = 0.95—reduced from 1.0 to account for normal production tolerances and variability.
    b. Shear and radial tension = 0.90—introduced in the new version of ASTM C361. A comparison can be made with the American Association of State Highway and Transportation Officials Load and Resistance Factor Design (AASHTO LRFD) capacity reduction factor for shear.
11. Allowable steel stress for hydrostatic head only, unchanged. Head is still measured to the centerline of the pipe.
12. Elliptical reinforcement, unchanged.
13. Concrete tensile stress = (0.433 * hydrostatic head * ID)/(two times wall thickness) $\leq 4.5\sqrt{f'c}$—upper limit changed to be $4.5\sqrt{f'c}$ because the standard now allows multiple concrete strengths (5 ksi or 6 ksi) in design; thus the maximum allowable concrete tensile stress should be variable. The prior versions of the standard assumed $4.8\sqrt{f'c}$ was less conservative than the $4.5\sqrt{f'c}$ that is in the current version.

14. Rational design procedures:
    a. Crack width control—following procedures of Section 12.10.4.2.4d of AASHTO LRFD [5]:
        i. Compressive thrust considerations
        ii. Tensile thrust considerations
    b. Shear capacity—following procedures described by Heger [6] but modified to introduce the F_{vp} term to account for impact of process and material control, which typically is 1.0 unless a manufacturer provides additional data to support increasing this factor to be greater than 1.0. When stirrups are required, follow design procedure illustrated in the American Society of Civil Engineers standard, ASCE 15-98 [7].
    c. Maximum flexural reinforcement:
        i. Limited by radial tension—following procedures described by Heger [6] but modified to introduce the F_{rp} term to account for impact of process and material control, which typically is 1.0 unless a manufacturer provides additional data to support increasing this factor to be greater than 1.0.
        ii. Limited by concrete compression—following procedures described by Heger [6].

The results of the changes described here are summarized in **Table 1**. Single values indicate a single cage. Values listed as XX/XX indicate that an inner and outer cage are required for the design.

As shown in **Table 1**, the steel areas highlighted in gray indicate how the standard changed only slightly from the pre-2011 design procedures to the current version of the standard when compared to ASTM C361, Table 1, values. But, as is shown, the steel areas provided in ASTM C361, Table 2 (last column), are somewhat reduced based on the improvement in the design realized by higher yield strength in the reinforcing steel.

Table 1 also illustrates that, for larger sizes with the deeper fills (D fills in the examples shown), shear reinforcement is required by the design, which was not required in the previous version. This trend matches with comparable fill heights requiring class (CL) CL4 up to 78 in. or CL5 up to 144 in. (refer to American Concrete Pipe Association Fill Height Tables [8]) for Type 3 bedding and resulting reinforcement designs not provided by ASTM C76 [9] because the designs are shear controlled for earth fills without internal pressure.

The steel areas for ASTM C361 and the USBR M-1 standards were in agreement up to their 1991 and 1992 versions as is shown by the first three columns of steel areas in **Table 1**. Changes to the USBR M-1 standard will be discussed in the following section.

USBR M-1

The USBR M-1 standard was published in 1969, 1974, 1984, and 1991. Example steel areas from the 1974 version and the 1991 version are included in **Table 1** illustrating a change in design philosophy taking place prior to the 1991 publication.

Design criteria were not included in the 1974 version of the USBR M-1 standard. A companion document was prepared by Sailer and Olander [10] in 1968 that included the design criteria similar to the early version of ASTM C361 as well as the reinforcing tables used in the 1974 version of the USBR M-1 standard. The 1991 version did include design criteria as an appendix.

Based on the document prepared by Sailer and Olander, and the agreement between the steel areas published by ASTM C361 and the USBR M-1 standards prior to 1990, the design criteria are assumed to have been the same. The 1991 version of the USBR M-1 standard includes higher steel areas than the previous version and incorporates the following assumptions:

1. Tabulated steel areas—single set of assumptions.
2. Required concrete compressive strength = 4,500 psi.
3. Required reinforcement yield strength = 40,000 psi.
4. Bedding angle:
   a. Pipe dead load = line bearing (0° bedding)
   b. All other loads = 90°
5. Bedding detail not included—the standard conservatively assumes that pipe will be installed such that the final bedding angle will be at least 90°.
6. Stress analysis determined using methods developed by Olander [4].
7. Loading combination for ultimate strength design:
   a. Pipe weight + earth load + internal water load + internal hydrostatic head
8. Unit weight of earth fill = 120 pcf + 24 pcf * (fill height/OD), 180 pcf max.
9. Load factor = 1.8 (all loads/limit states).
10. Capacity reduction factor = 1.0 (all loads/limit states). Standard specifically states that stirrups are not to be used even though some references require stirrups for some pipe classes.
11. Allowable steel stress for hydrostatic head only = 17,000 psi − 35 * hydrostatic head. Head is measured to the centerline of the pipe.
12. Elliptical reinforcement = 1.6 times steel area required for hydrostatic head.
13. General design of reinforced concrete is specified according to the AASHTO standard code [11] and the American Concrete Institute's ACI-318 [12] specifications. The year of these references is not stated in the USBR M-1 standard but should be assumed as the versions currently in publication at the time of the USBR M-1 publication.

Based on the design criteria presented in the appendix of the latest USBR M-1 publication, the major differences appear to be the unit weight of the earth fill and the bedding angle assumptions for pipe dead load.

Joint Standards

Joint design requirements and details are provided in the USBR M-1 standard, including a table indicating the minimum reinforcement to include in a concrete bell and the design procedure used to generate the table. The design of concrete bell reinforcement includes the following:

1. Hydrostatic head acts on a portion of the bell for an assumed length of 1.75 in. over the entire circumference of the bell.
2. The rubber gasket is assumed to exert 200 lb of force per inch over the entire inner perimeter of the bell.
3. The allowable tensile stress of the bell reinforcement is the same as assumed in the pipe wall design (17,000 psi − 35 * hydrostatic head).

These assumptions do not account for any additional strain that develops in the bell from external loads or deflection of the pipe in service or any additional load caused by joint shear, but the standard has performed well in the field. The standard also provides four different joint detail options including R1 (double spigot with steel bell band), R2 (integral cast steel end rings), R3 (concrete bell and spigot with gasket shoulders), and R4 (concrete bell and spigot with gasket groove). ASTM C361 does not provide guidance on joint design or details. Both standards provide guidance on gasket materials and properties.

The USBR has also developed two computer programs:

1. RCPIPE9.FOR: FORTRAN code for design of the pipe wall according to USBR M-1 procedures.
2. RCBELL1.FOR: FORTRAN code for design of the concrete bell reinforcement according to USBR M-1 procedures.

According to the USBR M-1 document, these programs are available by contacting the USBR, ATTN Code 86-68150, PO Box 25007, Denver, CO, 80225.

During review of the USBR M-1 standard, one can notice that the appendix states that the internal hydrostatic pressure is measured from the inside top of the pipe to the design hydraulic gradient. But the pressure classes indicate that the pressure is to the centerline of the pipe. It should be noted that the pressure at the top of a pipe will be less than the pressure at the bottom of the pipe, with the average pressure found at the centerline of the pipe. Olander stress analysis accounts for the weight of the water inside the pipe for the structural design of the pipe wall. This difference may be relatively negligible for most pipe sizes but could have significant impact on a larger diameter pipe where the gasket at the bottom of the pipe will be under significantly higher pressure than the gasket at the top of the pipe. Designers should consider this when analyzing the gasket for sealing pressure.

Summary

This document presents a portion of the history behind the development of the ASTM C361 and USBR M-1 standards. The documents initially followed similar paths in the 1950s and 1960s and then diverged in the 1990s. Since that time, extensive discussion within the ASTM C13.04 committee has led to the development of new design criteria for the current ASTM C361 tables. The USBR has retired the M-1 standard and adopted ASTM C361 as the project standard and has incorporated the joint design details into standard plates for insertion into their plans. The ASTM C13.04 subcommittee has discussed the addition of these details to the ASTM C361 standard, but no decision had been made as of the date of this paper.

The current version of ASTM C361 incorporates a comprehensive collection of limit states and rational design checks that are current with industry state of the art. These procedures allow a designer or specifier the flexibility to take advantage of current materials available that improve the standard designs and to prepare special designs for sizes or classes not covered by the tables with full confidence.

References

[1] ASTM C361, *Standard Specification for Reinforced Concrete Low-Head Pressure Pipe*, ASTM International, West Conshohocken, PA, 2014, www.astm.org

[2] USBR M-1, *Standard Specifications for Reinforced Concrete Pressure Pipe*, U.S. Bureau of Reclamation, Washington, DC, 1991.

[3] AWWA M9, *Concrete Pressure Pipe—Manual of Water Supply Practices*, 3rd ed., American Water Works Association, Washington, DC, 2008.

[4] Olander, H. C., *Stress Analysis of Concrete Pipe, Engineering Monograph No. 6*, U.S. Bureau of Reclamation, Washington, DC, 1950.

[5] *AASHTO, LRFD Bridge Design Specifications*, 7th ed., American Association of State Highway and Transportation Officials, Washington, DC, 2014.

[6] Heger, F. J., "SIDD Installation and Direct Design for Reinforced Concrete Low-Head Pressure Pipe," *Concrete Pipe for the New Millennium, ASTM STP1368*, I. I. Kaspar, and J. I. Enyart, eds., ASTM International, West Conshohocken, PA, 2000, http://dx.doi.org/10.1520/STP14995S

[7] ASCE 15-98, *Standard Practice for Direct Design of Buried Precast Concrete Pipe Using Standard Installations (SIDD)*, American Society of Civil Engineers, New York, 2001.

[8] ACPA, *LRFD Fill Height Tables for Concrete Pipe*, American Concrete Pipe Association, Irving, TX, 2013, www.concretepipe.org

[9] ASTM C76, *Standard Specification for Reinforced Concrete Culvert, Storm Drain, and Sewer Pipe*, ASTM International, West Conshohocken, PA, 2015, www.astm.org

[10] Sailer, R. and Olander, H. C., "*Design of Reinforced Concrete Low Head Pressure Pipe—Table 1*," U.S. Bureau of Reclamation, Washington, DC, 1968.

[11] *AASHTO, Standard Specifications for Highway Bridges*, Section 17, American Association of State Highway and Transportation Officials, Washington, DC.

[12] ACI-318, *Building Code Requirements for Reinforced Concrete*, American Concrete Institute, Detroit, MI.

STP 1601, 2017 / available online at www.astm.org / doi: 10.1520/STP160120160129

Steven J. DelloRusso[1] and George Wayne Hodge[2]

History of the ASTM Specifications for Precast Reinforced Concrete Box Culvert Sections

Citation

DelloRusso, S. J. and Hodge, G. W., "History of the ASTM Specifications for Precast Reinforced Concrete Box Culvert Sections," *Concrete Pipe and Box Culverts, ASTM STP1601*, J. J. Meyer and J. Beakley Eds., ASTM International, West Conshohocken, PA, 2017, pp. 59–74, http://dx.doi.org/10.1520/STP160120160129[3]

ABSTRACT

Since about 1974, ASTM has published standard specifications for precast reinforced concrete box culvert sections. These standard specifications included tabulated reinforcing designs for a variety of sizes and earth cover depths, as well as manufacturing requirements and procedures for quality control. This paper chronicles the history and development of these ASTM standards published under the jurisdiction of ASTM C13 Committee on Concrete Pipe, with a focus on the specific methods used to develop the reinforcing tables and changes to the specifications with time to reflect industry consideration of design code changes. We provide details behind the withdrawal of the older specifications and the creation and passing of newer standards for box culvert sections. This paper also examines the history of the ASTM specifications with respect to AASHTO materials standards and design codes.

Keywords

box culvert, ASTM, specification history, BOXCAR, design, reinforced concrete box culvert, standard designs

Manuscript received September 30, 2016; accepted for publication November 15, 2016.

[1]Simpson Gumpertz & Heger, Inc., 41 Seyon St., Waltham, MA 02453

[2]Rinker Materials Concrete Pipe Division, 6560 Langfield Rd., Bldg. 3, Houston, TX 77092

[3]ASTM Symposium on *Concrete Pipe and Box Culverts* on December 7, 2016 in Orlando, Florida.

Introduction

In the early 1970s, there were initiatives to develop standards for the manufacture and design of precast reinforced concrete box culverts (RCBs). RCBs had been used for many years prior and were mostly cast-in-place. Precast offered the benefits of plant manufacture, which is conducive to strict quality control measures and inspection, and the finished product could be installed by rapid cut and cover installation techniques, similar to round concrete pipe. These initiatives resulted in the publication and acceptance of standards for precast RCBs starting in 1974, and these have continued to evolve to the present.

Initial Development and Computerized Design

A cooperative venture to develop a manufacturing specification, including standard designs for precast RCBs, was started in 1971 by the Virginia Department of Highways and the American Concrete Pipe Association (ACPA) and funded by the Wire Reinforcement Institute [1]. This venture led to the development of a computer program for the structural design of box sections that allowed for variations in the geometry of the structure including span length and rise height as well as top and bottom slab and wall thicknesses. Because material properties such as concrete and steel reinforcing strengths differ for precast versus cast-in-place, variables were incorporated into the program to accommodate these differences.

The details of the program and its design methods are documented by Latona, Heger, and Healey [1]. The program was used to develop standard sizes based on input from manufacturers and end users. Standard designs assumed a uniform thickness of the walls, slabs, and vertical and horizontal haunch dimensions.

Some fundamental characteristics of the reinforcing designs of the initial tables included the following:

- Structural analysis used the stiffness matrix method, and a 1-ft section of box culvert was analyzed as a four-member frame as shown in **Fig. 1**.
- Design loads considered the following, as depicted in **Fig. 2**:
  - Culvert self-weight
  - Vertical and horizontal soil loads
  - Internal fluid
  - Live loads
- Reinforcing design was per the ultimate strength design methods, current at the time of the program development, for combined bending and axial load.
- Reinforcing areas for each member were calculated based on factored ultimate loads and serviceability.
- The program did not design shear reinforcing, but the output noted if shear reinforcing was needed. Shear did not control any of the tabulated designs.
- The original program had the following limitations:
  - Only single cell culverts.
  - Burial depth range: 2.0 ft to 100 ft.

FIG. 1 Frame model used for structural analysis.

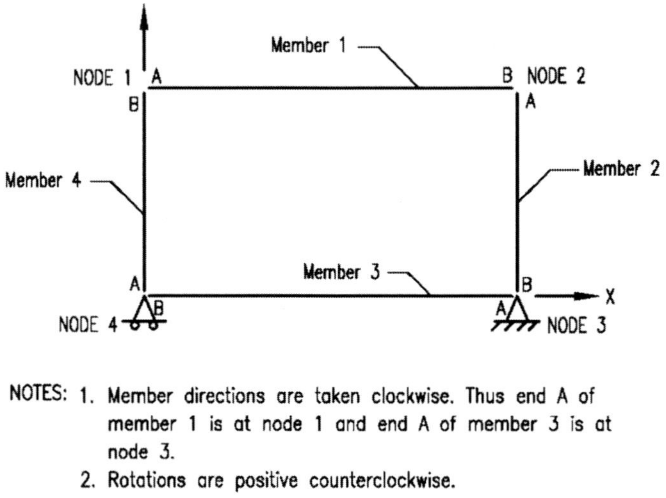

- Span range: 3.0 ft to 12.0 ft.
- Rise range: 2.0 ft to 12.0 ft.
- Loads were limited to the aforementioned design loads.

The initial standard tables were incorporated into the first issue of ASTM C789, *Standard Specification for Precast Reinforced Concrete Box Sections for Culverts, Storm Drains, and Sewers* [2]. After some modifications to the design program including distribution of wheel loads at the surface of the box and limits of total service load steel stress and service live load steel stress, or fatigue, new standard design tables for low earth cover heights and updated ASTM C789 standard tables were created [3].

The computer program evolved over time and was publically released as a microcomputer program named BOXCAR. The program was developed by Simpson Gumpertz & Heger Inc. (SGH) in cooperation with the Federal Highway Administration (FHWA) and the ACPA as part of an FHWA project to develop standard designs for improved inlets. It is further detailed in the report *Structural Design Manual for Improved Inlets and Culverts* [4].

The publically released BOXCAR V1.x program consisted of analysis routines written in the original FORTRAN program coupled with a terminal user interface for data input and for post-processing and viewing the program output. The program required the MS-DOS operating system.

There were some differences between the publically released version of BOXCAR and the program used to create the ASTM standard tables. The differences

FIG. 2 Loads in BOXCAR program. Note the original program did not include approaching truck loads, variable lateral soil loads, or vertical or horizontal surcharge loads.

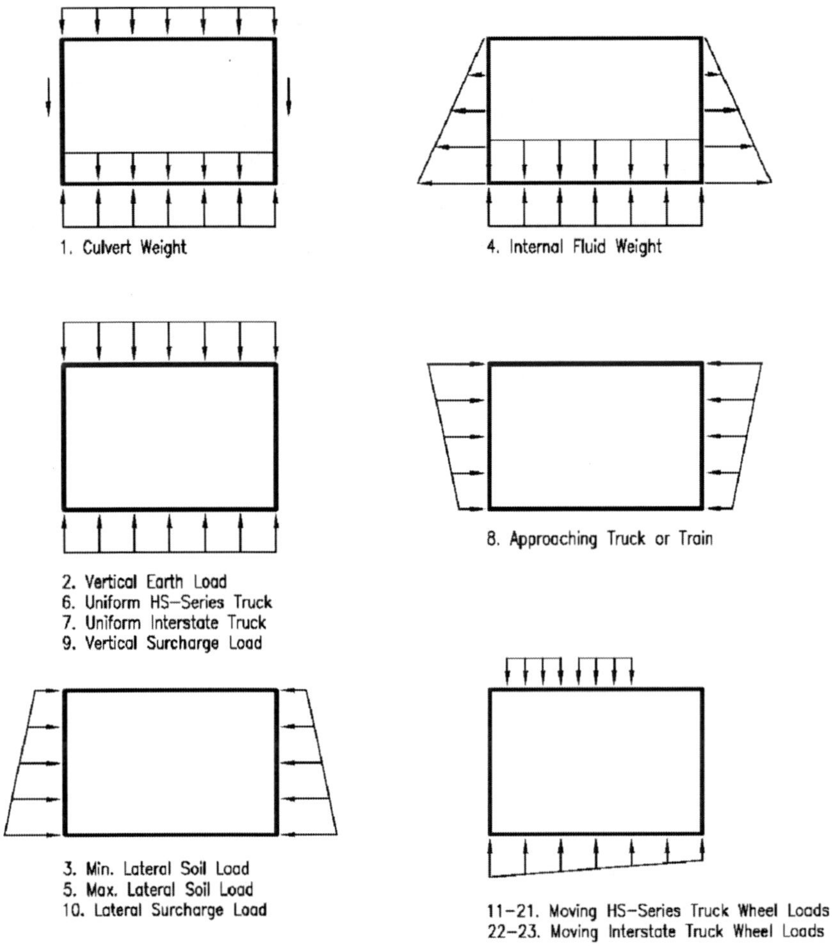

included some live load assumptions and some reinforcing design changes (i.e., design is in accordance with the American Association of State Highway and Transportation Officials [AASHTO] as opposed to ACI). The details of these differences were explained in the original BOXCAR manual [5].

In 2000, BOXCAR was rewritten by SGH for the ACPA in MS Visual Basic as Version 2.x to incorporate code updates and to run under the MS-Windows operating system. BOXCAR Version 2.x allowed the user to select either the AASHTO standard bridge design code or the newly adopted AASHTO Load and Resistance Factor (LRFD) bridge design code.

After a subsequent rewrite in 2010, BOXCAR Version 3.x was issued to incorporate numerous changes and revisions to the LRFD code. BOXCAR 3.x was in compliance with AASHTO 2007 at the time of its release.

Fig. 2 and Fig. 3 show the general load conditions and design sections used in the BOXCAR culvert design programs.

Fig. 4 shows the general configuration of the box section reinforcing and the nomenclature used for the steel reinforcing areas in both the box culvert designs standards and the BOXCAR design programs.

ASTM

All of the ASTM box culvert standards include Appendix X1, which lists the design assumptions used to generate the standard tables. Table 1 and Table 2 summarize these listed assumptions for culvert design with earth cover heights greater than 2 ft and with less than 2 ft, respectively. Only revisions with changes to this criteria or where the

FIG. 3 Design locations from the BOXCAR program. The original program used a similar configuration.

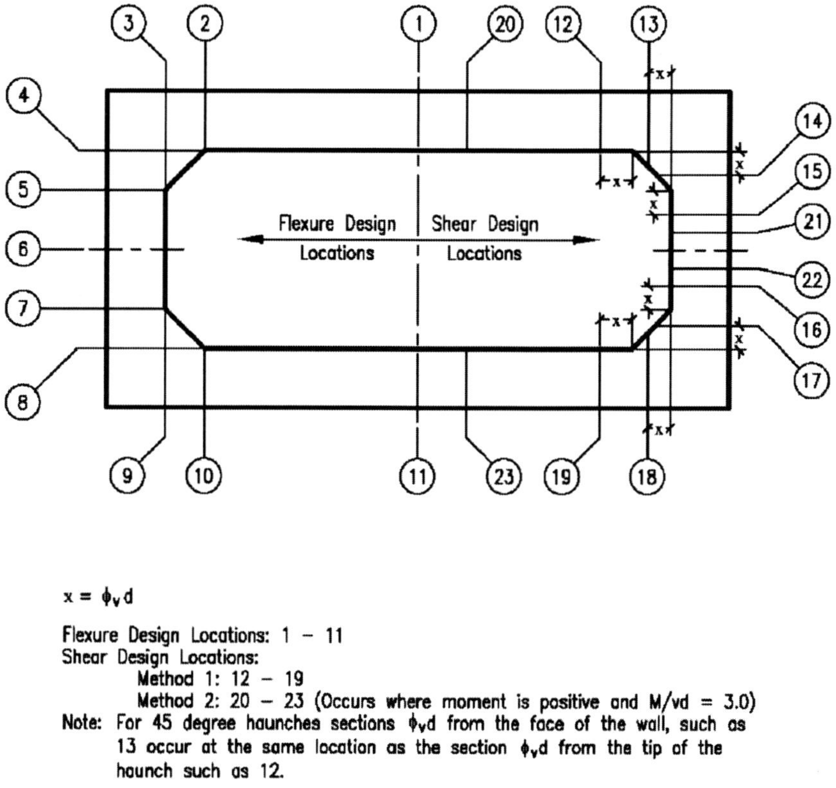

FIG. 4 Typical box culvert reinforcing configuration and nomenclature used in the box culvert standards and design programs. The reinforcing designations are as follows: As1 is the sidewall outside face reinforcing; As2 is the top slab inside face reinforcing; As3 is the bottom slab inside face reinforcing; As4 is the sidewall inside face reinforcing; As5 is the bottom distribution reinforcing; As6 is the top distribution reinforcing; As7 is the top slab outside face reinforcing; and As8 is the bottom slab outside face reinforcing.

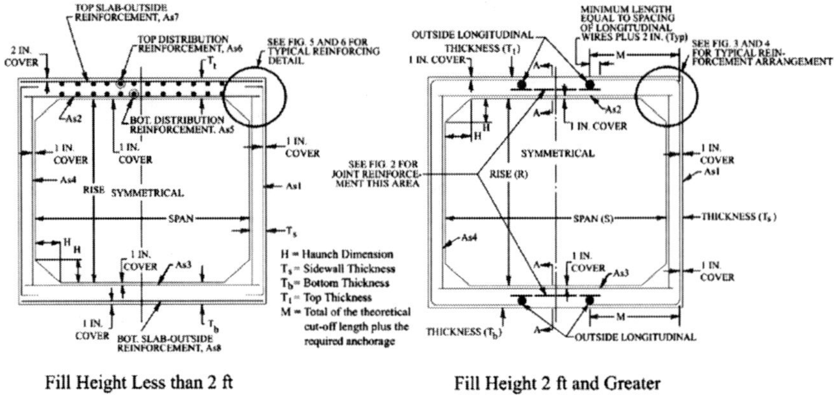

design tables have been revised are included in this comparison. The sections that follow review each standard and discuss table revision history for each standard.

Although there are companion metric (SI) specifications to ASTM C789, ASTM C850, *Standard Specification for Precast Reinforced Concrete Box Sections for Culverts, Storm Drains, and Sewers with Less Than 2 ft of Cover Subjected to Highway Loadings* [6], and to ASTM C1433, *Standard Specification for Precast Reinforced Concrete Monolithic Box Sections for Culverts, Storm Drains, and Sewers* [7], the discussions in this paper focus only on the Imperial unit specifications. There were no separate analyses performed for the SI specifications; instead, the SI design tables were generated by conversion of the Imperial unit tables.

ASTM C789

ASTM C789 was first adopted in 1974 by ASTM Committee C13 on Concrete Pipe. It contains tabular designs for various combinations of span times rise ranging from 3 ft by 2 ft to 10 ft by 10 ft and for three loading conditions: combined earth dead load and AASHTO HS20 live load; combined earth dead load and AASHTO interstate live load (when the interstate live load is controlled); or dead load only (with no live load).

The tables presented in the standard considered only box culverts with 2 ft or more of soil cover. The live load application to the buried top slab assumed the

TABLE 1 Comparison of design parameters for ASTM standards for box culverts with greater than 2 ft of fill.

	C789-74	C789-76 Through C789-95a	C1433 Designs with Greater than 2 ft of Earth	C1577 Designs with Greater than 2 ft of Earth
Referenced AASHTO Design Specification	Standard Specification 10th ed. 1969	Standard Specification 12th ed. 1973	Standard Specification, 1997	LRFD Specification, 2004 w/2005 Interim
Material Properties:				
Welded wire fabric, minimum specified yield stress	65,000 psi	65,000 psi	65,000 psi	65,000 psi
Concrete, minimum specified compressive strength	5,000 psi	5,000 psi	5,000 psi	5,000 psi
Soil Data:				
Unit weight	120 lb/ft′	120 lb/ft′	120 lb/ft′	120 lb/ft′
Ratio of lateral to vertical pressure from weight of earth	0.33	0.25 min. to 0.50 max	0.25 min. to 0.50 max	0.25 min. to 0.50 max
Additional lateral pressure from approaching truck wheels		$700 \div H$, lb/ft^2 where H = earth cover, ft	$700 \div H$, lb/ft^2 where H = earth cover, ft	Lateral Live Load Pressure: From 0 to 5 ft 160 psf
External water table	Below box section invert	Below box section invert	Below box section invert	Below box section invert
Effective weight coefficient	1.0	1.0	1.15 (SSIF)	$F_e = 1 + 0.20$ (H/Bc) B_c = outside width of box $F_e$max = 1.15
Capacity Reduction Factors	From ACI 318-71*	From ACI 318-71*	From AASHTO	From AASHTO
Shear	0.85	0.85	0.90	0.9
Axial compression combined with bending	0.70 to 0.90	0.70 to 0.90	0.95	1.0
Loading Data:				
Load factor—dead load	1.5	1.5	1.3	1.3 plus 1.05 load modifier
Load factor—live load	2.2	2.2	2.2	1.75 plus 1.2 multiple presence factor
Truck axle load:				
H_2O (Table 1)	32,000 lbf	32,000 lbf	32,000 lbf	HL93 live load checks both conditions

TABLE 1 *(Continued)*

	C789-74	C789-76 Through C789-95a	C1433 Designs with Greater than 2 ft of Earth	C1577 Designs with Greater than 2 ft of Earth
Interstate (Table 2)	2 @ 24,000 1bf each	2 @ 24,000 1bf each	2 @ 24,000 1bf each	
None (Table 3)				
Impact (variable with depth) see AASHTO Bridge Specifications, 1969	0 to 20 %	0 to 20 %	0 to 20 %	0.3 * (1–0.125H)
Live load distribution:				
Parallel to span	Point load at 1.75H	Point load at 1.75H	Point load at 1.75H	(10 + 1.15H) in.
Perpendicular to span	Point load at 1.75H – 5 ft max per wheel	Point load at 1.75H – 5 ft max per wheel	Point load at 1.75H – 6 ft max per wheel	(20 + 1.15H) in.
Uniform internal pressure	0.0	0.0	0.0	0.0
Depth of water in box section	Equal to inside height	Equal to inside height	Equal to inside height	Equal to inside height
External ground water pressure	0.0	0.0	0.0	0.0
crack control	Limit spacing to 4 in.		Used Heger/ McGrath method –04 used Gergley-Lutz (AASHTO)	Per LRFD code
Structural Arrangement:				
Concrete cover over steel	1.0 in.	1.0 in.	1.0 in.	1.0 in.
Slab thickness	1/12 times inside span plus 1.0 in. up to 7-ft span—1/12 times inside span above 7-ft span	1/12 times inside span plus 1.0 in. up to 7-ft span—1/12 times inside span above 7-ft span	1/12 times inside span plus 1.0 in. up to 7-ft span—1/12 times inside span above 7-ft span	1/12 times inside span plus 1.0 in. up to 7-ft span—1/12 times inside span above 7-ft span
Side wall thickness	Equal to slab thickness	Equal to slab thickness	Equal to slab thickness	Equal to slab thickness
Haunch dimensions	Vertical and horizontal dimensions both equal to slab thickness	vertical and horizontal dimensions both equal to slab thickness	Vertical and horizontal dimensions both equal to slab thickness	Vertical and horizontal dimensions both equal to slab thickness
Minimum reinforcing inside face slabs and side walls, outside face side walls and corners of slabs	0.002 bt	0.002 bt	0.002 bt	0.002 bt

TABLE 2 Comparison of design parameters for ASTM standards for box culverts with less than 2 ft of fill.

	C850-76	C850-95a	C1433 Designs with Less than 2 ft of Earth	C1577 Designs with Less than 2 ft of Earth
Referenced AASHTO Design Specification	Standard Specification 10th Ed. 1969	Standard Specification 12th Ed. 1977	Standard Specification, 1997	LRFD Specification, 2004 w/2005 Interim
Material Properties:				
Welded wire fabric, minimum specified yield stress	60,000 psi	60,000 psi	65,000 psi	65,000 psi
Deformed bars (longitudinal distribution reinforcement)	60,000 psi	60,000 psi	60,000 psi	
Concrete, minimum specified compressive strength	5,000 psi	5,000 psi	5,000 psi	5,000 psi
Soil Data:				
Unit weight	120 lb/ft′	120 lb/ft′	120 lb/ft′	120 lb/ft′
Ratio of lateral to vertical pressure from weight of earth	0.25 min. to 0.50 max	0.25 min. to 0.50 max	0.25 min. to 0.50 max	0.25 min. to 0.50 max
Additional lateral pressure from approaching truck wheels	800 lbs/ft^2 to 1 ft earth cover $700 \div H$, lb/ft^2 where H = depth of earth cover, ft when depth exceeds 1 ft	800 lbs/ft^2 to 1 ft earth cover $700 \div H$, lb/ft^2 where H = depth of earth cover, ft when depth exceeds 1 ft	800 lbs/ft^2 to 1 ft earth cover $700 \div H$, lb/ft^2 where H = depth of earth cover, ft when depth exceeds 1 ft	Lateral Live Load Pressure: From 0 to 5 ft 160 psf $5 \geq 10$ft $160 - [(H-5)/(10-5)](160-120)$ psf $10 \geq 20$ ft $120 - [(H-10)/(20-10)](120-80)$ psf 20 ft or greater 80 psf
External water table	Below box section invert	Below box section invert	Below box section invert	Below box section invert
Effective weight coefficient	1.0	1.0	1.15 (SSIF)	$F_e = 1 + 0.20$ (H/Bc) B_c = outside width of box $F_e$max = 1.15
Capacity Reduction Factor	from ACI 318-71*	from ACI 318	from AASHTO	from AASHTO
Shear	0.85	0.85	0.90	0.9
Axial compression combined with bending	0.70 to 0.90	0.70 to 0.90	0.95	1.0

TABLE 2 *(Continued)*

	C850-76	C850-95a	C1433 Designs with Less than 2 ft of Earth	C1577 Designs with Less than 2 ft of Earth
Loading Data:				
Load factor—dead load	1.5	1.5	1.3	1.3 plus 1.05 load modifier
Load factor—live load	2.2	2.2	2.2	1.75 plus 1.2 multiple presence factor
Truck axle load:				
H20 (Table 1)	32,000 lbf	32,000 lbf	32,000 lbf	HL93 live load checks both conditions
Interstate (Table 2)	2 @ 24,000 1bf each	2 @ 24,000 1bf each	2 @ 24,000 1bf each	
None (Table 3)				
Impact (variable with depth) see AASHTO Bridge Specifications, 1969	30 % to 20 %	30 % to 20 %	0 to 30 %	0.3*(1–0.125H)
Live load distribution:				
Parallel to span	(8 + 1.75H) in.	(8 + 1.75H) in.	Concentrated point load	(10 + 1.15H) in.
Length of box section resisting wheel load	48 + 0.06*(span-haunch)	48 + 0.06*(span-haunch)	48 + 0.06*(span-haunch)	96 + 1,44 span (ft) for axle
Uniform internal pressure	0.0	0.0	0.0	0.0
Depth of water in box section	Equal to inside height	Equal to inside height	Equal to inside height	Equal to inside height
External ground water pressure	0.0	0.0	0.0	0.0
Fatigue	Live load stress limit = 21 ksi	Live load stress limit = 21 ksi	Live load stress limit = 21 ksi	Live load stress limit = 21 ksi
Service load stress limit	Total service load stress limit = 36 ksi	Total service load stress limit = 36 ksi	Total service load stress limit = 36 ksi	
Crack width control	Limit spacing to 4 in.	Limit spacing to 4 in.	Used Heger/ McGrath method –04 used Gergley-Lutz (AASHTO)	Per LRFD code
Structural Arrangement:				
Concrete cover over steel				
top of top slab	2.0 in.	2.0 in.	2.0 in.	2.0 in.
all other surfaces	1.0 in.	1.0 in.	1.0 in.	1.0 in.

TABLE 2 *(Continued)*

	C850-76	C850-95a	C1433 Designs with Less than 2 ft of Earth	C1577 Designs with Less than 2 ft of Earth
Slab thickness	Varies - See tables	Varies - See tables	1/12 times inside span plus 1.0 in. up to 7-ft span—1/12 times inside span above 7-ft span	1/12 times inside span plus 1.0 in. up to 7-ft span—1/12 times inside span above 7-ft span
Side wall thickness	Varies - See tables	Varies - See tables	Equal to slab thickness	Equal to slab thickness
Haunch dimensions	Vertical and horizontal dimensions both equal to side wall thickness	Vertical and horizontal dimensions both equal to slab thickness	Vertical and horizontal dimensions both equal to slab thickness	Vertical and horizontal dimensions both equal to slab thickness
Minimum reinforcing inside face slabs and side walls, outside face side walls and corners of slabs	0.002 bt or 0.125 in^2/ft whichever is greatest	0.002 bt or 0.125 in^2/ft whichever is greatest	0.002 bt	0.002 bt

truck wheel loads as point loads that were distributed through the soil to a square with sides equal to 1.75 H.

Original designs were in accordance with the tenth edition of the *AASHTO Standard Specifications* (1969). The designs assumed that the ratio of lateral soil pressure to vertical soil load was 0.33 and did not consider any lateral load from approaching trucks. Based on the design results, no steel reinforcing was required on the inside face of the sidewalls (As4). Therefore, the design tables only listed reinforcing areas for As1, As2, and As3.

A subsequent edition, ASTM C789-76, updated the referenced design specification to the *AASHTO Standard Specifications* (1973) and assumed that the lateral soil pressure ratio varied from 0.25 to 0.50. This edition also included load from approaching trucks. The revised design tables require As4 reinforcing.

For the ASTM C789-81 edition, the design tables were expanded to a maximum 12 ft span by 12 ft rise, but the design method and assumptions were unchanged through the final approved version, ASTM C789-95a.

The ASTM C789 standard was retired and superseded by ASTM C1433 in 2000.

ASTM C850

Standard tables for box culverts with less than 2 ft of fill were first published in 1976 as a new and separate specification, ASTM C850. It referenced *AASHTO Standard Specifications* (1973 with 1974 interims) as the design specification. Based on a consensus of structural designers, manufacturers, and transportation officials, enhancements were incorporated into the original design program to distribute the live load at earth fill heights from 0 ft to 2 ft. Subsequently, companion standard design tables were created for the various culvert span times rise geometries ranging from 3 ft by 2 ft to 12 ft by 12 ft for the lower fill heights.

Due to the magnitude of the live load at these lower fill heights, the ASTM C850 designs typically had a thicker top slab. Additionally, the nominal concrete cover over the top slab outside face was increased from 1 in. to 2 in. for improved durability, and the designs required longitudinally oriented distribution steel in the top slab to mitigate the effects of the concentrated live load.

The referenced design specification for ASTM C850 was updated in 1995 to *AASHTO Standard Specifications* (1977). ASTM C850 remained unchanged from 1995 until it was retired and superseded by ASTM C1433 in 2000.

ASTM C1433

Prompted by changes in the AASHTO standard code provisions, ASTM Committee C13 decided to update the design tables of ASTM C789 and ASTM C850. With this update, the committee combined the two existing standards into one standard and eliminated separate standards for shallow earth covers less than 2 ft and for greater than 2 ft of earth cover. The new tables were created using the existing BOXCAR program with some modifications for conformance to the AASHTO code provisions. ASTM C1433 was first adopted in 1999 and published in 2000. The new standard eliminated the earth load only design tables, and had tables only for the HS20 live load and interstate live load.

Note that the tables in ASTM C1433 maintain a distinction in the basic box culvert geometry for designs for 2 ft and below, consistent with the ASTM C850 geometry, requiring: thicker top and bottom slabs as necessary, greater concrete cover over the top slab outside face reinforcing, and distribution steel in the top slab.

For example, **Table 3** presents a comparison of the steel reinforcing areas for an 8 ft by 6 ft box culvert as reported in the latest ASTM C850/ASTM C789 standards to revisions of the ASTM C1433 standard. This comparison finds that, for the first issue of ASTM C1433, sidewall outside face reinforcing steel (As1) and top slab inside face reinforcing steel (As2) were noticeably reduced for the box under 10 ft of fill and that the top slab inside face reinforcing (As2) increased for the new tables at 0 ft to 2 ft. The ASTM C1433 tables included the effects of thrust in the sidewalls, which had an impact on the As1 reinforcing. It is unclear what code/programming influences resulted in the changes to the As2 reinforcing, but the subsequent revision of the standard in 2004 resulted in steel areas more consistent with the previous ASTM C850-95a standard.

Published tables in the ASTM C1433-00 and ASTM C1433-01 standards contained an error in the tabulated extension of the sidewall outside steel into the top and bottom slabs, or the "M" dimension. Typical precast construction utilizes two

TABLE 3 Comparison of reinforcing steel areas (in^2/ft) for an 8 ft x 6 ft box culvert.

Standard	Earth Cover	Live Load	Reinforcing Steel Designation As1	As2	As3	As4	M, in.
C789-95a	10	HS20	0.27	0.37	0.39	0.19	27
C1433-01	10	HS20	0.21	0.32	0.34	0.19	18
C1433-02a	10	HS20	0.21	0.32	0.34	0.19	45
C1433-04	10	HS20	0.22	0.35	0.37	0.19	45
C1577-05	10	HL-93	0.22	0.33	0.34	0.19	45

Standard	Earth Cover	Live Load	Reinforcing Steel Designation As1	As2	As3	As4	As7	As8	As5	As6
C850-95a	0-2	HS20	0.32	0.56	0.33	0.19	0.25	0.19	0.21	0.19
C1433-02a	0-2	HS20	0.28	0.64	0.33	0.19	0.19	0.19	0.24	0.19
C1433-04	0-2	HS20	0.26	0.59	0.28	0.19	0.19	0.19	0.22	0.19*
C1577-05	0-2	HL-93	0.22	0.42	0.35	0.19	0.19	0.19	0.19	0.19*

*As6 was eliminated in the 2008 revision of C1433 and the 2014a revision of C1577.

full reinforcing cages (one inside and one outside), so the error likely did not have an effect on product manufactured at the time, but the tables were revised in the 2002 revision.

All design tables for ASTM C1433 were revised by SGH and the ACPA in 2004 and published in 2005 using an updated BOXCAR program, Version 2.03. The updated program was a rewrite to support the Windows operating system; it added more design sections for a more refined analysis and incorporated code changes for live load application and design parameters. Some of these changes are shown in **Table 1** and **Table 2**, which compare the design appendices of the standards.

Based on the research [6] finding that any compression distribution steel contribution to the distribution of the live load or the performance of the box section is negligible, the ASTM C1433-08 standard eliminated the requirement for top slab outside face distribution reinforcement, As6.

ASTM C1577

In 2005, ASTM Committee C13 issued a new standard, ASTM C1577, *Standard Specification for Precast Reinforced Concrete Monolithic Box Sections for Culverts, Storm Drains, and Sewers Designed According to AASHTO Load and Resistance Factor (LRFD) Bridge Design Specifications* [8]. A comparison of the differences between the AASHTO standard code and the AASHTO LRFD code is beyond the scope of this paper but, as described in [9], the LRFD approach is consistent with the ultimate strength design followed for the standard code box culvert tables.

However, load factors and load modifiers are expanded in the LRFD code to account for variability of the load, importance of the structure, and so on. From an analysis standpoint, this requires a different assembling of the design forces for the reinforcing design equations and evaluation of limit states as opposed to the standard design code. BOXCAR Version 2.03, used to generate the original ASTM C1577 design tables, incorporates revisions for substantial compliance with the AASHTO LRFD *Bridge Design Specifications* (2004, with 2005 interims). Fundamental design parameters used to develop the tables are listed in Table 1 and Table 2.

The design live load for the ASTM C1577 tables is the AASHTO HL-93 design load, which checks both the design truck and the design tandem. The design truck assumes 32-k axles spaced at 14 ft, which is consistent with the AASHTO standard code-specified HS-20 truck. The design tandem is two 25-k axles spaced at 4 ft apart, which is consistent with the interstate loading of the AASHTO standard code, except the axle load is increased from 24 k to 25 k. Therefore, for practicable purposes, the table designs of ASTM C1577 can be compared with the governing steel areas from either the ASTM C1433, Table 1 (HS20) or Table 2 (interstate live load). Table values for ASTM C1577 are included in the example 8 ft by 6 ft reinforcing steel comparison in Table 3. In general, reinforcing areas were reduced due to the revisions in live load application and the crack control provisions in the LRFD code.

As the use of precast RCBs continued to grow, so did the number of sizes of RCBs that would be used on a routine basis. Special designs were required whenever nonstandard-sized RCBs were specified. To address this trend, ASTM C1577-13a provided 16 additional box sizes to the design table. The new box sizes provided for smaller rise dimensions than were available in the previous design tables.

Based on the research [10] finding that any compression distribution steel contribution to the distribution of the live load or to the performance of the box section is negligible, and consistent with ASTM C1433, the ASTM C1577-14a edition of the standard eliminated the requirement for top slab outside face distribution reinforcement, As6. This edition also expanded to the design tables once again to include seven additional span times rise combinations using the same program utilized to develop the original tables.

AASHTO M259 and AASHTO M273

AASHTO M259, *Standard Specification for Precast Reinforced Concrete Box Sections for Culverts, Storm Drains, and Sewers*, and AASHTO M273, *Standard Specification for Precast Reinforced Concrete Box Sections for Culverts, Storm Drains, and Sewers with Less Than 2 Ft of Cover Subjected to Highway Loadings*, are versions of the ASTM C789 and ASTM C850 standards that have been adopted by AASHTO, including the standard design tables. Both of the AASHTO standards are currently active as AASHTO M-259-11-UL and AASHTO M273-11-UL, respectively. The current version of AASHTO M259 references ASTM C789-95a, and AASHTO M273 references ASTM C850-95a. Both of the AASHTO standards reference the AASHTO standard bridge design code that was last revised in 2002. For new designs, AASHTO has superseded the *Standard Bridge Design Specifications* in favor of the *AASHTO LRFD Bridge Design Specifications.*

Presently, AASHTO has not adopted any box culvert design standards for the LRFD specifications but has added a note to both the AASHTO M259 and AASHTO M273 standards directing the user to ASTM C1577 if the box culvert needs to be designed in accordance with the LRFD design specification.

Future Revisions

Newer box culvert design programs have been developed that operate on the newer computer platforms and that incorporate the latest AASHTO code provisions into their design calculations. Therefore, there is an industry need to update the standard design tables. Because AASHTO is no longer updating the standard bridge design code, it is the authors' belief that only the standard design tables of ASTM C1577 will be revised within the next few years.

References

[1] Latona, R. W., Heger, F. J., and Healey, M., "Computerized Design of Precast Reinforced Concrete Box Culverts," *Hwy. Res. Rec.*, Issue 443, 1973, pp. 40–51.

[2] ASTM C789-00, *Standard Specification for Precast Reinforced Concrete Box Sections for Culverts, Storm Drains and Sewers,* ASTM International, West Conshohocken, PA, 2000, www.astm.org

[3] Heger, F. J. and Long, K. N., "Structural Design of Precast Concrete Box Sections for Zero to Deep Earth Cover Conditions and Surface Wheel Loads," *Concrete Pipe and the Soil-Structure System, ASTM STP630*, M. Bealey and J. D. Lemons, Eds., ASTM International, West Conshohocken, PA, 1977, pp. 3–16, http://dx.doi.org/10.1520/C1577-17

[4] McGrath, T. J. and Heger, F. J., *Structural Design Manual for Improved Inlets and Culverts*, FHWA-IP-83-6, FHWA, Turner-Fairbank Highway Research Center, McLean, VA, 1983.

[5] *BOXCAR Version 1.0—User's Manual, Appendix C: Differences in Analysis Between Boxcar and the Programs that Generated ASTM Standard Designs*, Federal Highway Administration Report FHWA-IP-89-019, April, 1989.

[6] ASTM C850-00, *Standard Specification for Precast Reinforced Concrete Box Sections for Culverts, Storm Drains, and Sewers with Less Than 2 ft of Cover Subjected to Highway Loadings*, ASTM International, West Conshohocken, PA, 2000, www.astm.org

[7] ASTM C1433-16b, *Standard Specification for Precast Reinforced Concrete Monolithic Box Sections for Culverts, Storm Drains, and Sewers*, ASTM International, West Conshohocken, PA, 2016, www.astm.org

[8] ASTM C1577-17, *Standard Specification for Precast Reinforced Concrete Box Sections for Culverts, Storm Drains, and Sewers Designed According to AASHTO Load and Resistance Factor (LRFD) Bridge Design Specifications*, ASTM International, West Conshohocken, PA, 2017, www.astm.org

[9] Rund, R. E. and McGrath, T. J., "Comparison of AASHTO Standard and LRFD Code Provisions for Buried Concrete Box Culverts," *Concrete Pipe for the New Millennium, ASTM STP1368*, J. I. Enyart and I. I. Kaspar, Eds., ASTM International, West Conshohocken, PA, 1999, pp. 45–60, http://dx.doi.org/10.1520/STP14997S

[10] Abolmaali, A. and Garg, A., "Effect of Wheel Live Load on Shear Behavior of Precast Reinforced Concrete Box Culverts," *J. Bridge Eng.*, Vol. 13, No. 1, 2008, pp. 93–99.

STP 1601, 2017 / available online at www.astm.org / doi: 10.1520/STP160120160119

Omid Gooranorimi,[1] John Myers,[2] and Antonio Nanni[3]

GFRP Reinforcements in Box Culvert Bridge: A Case Study After Two Decades of Service

Citation

Gooranorimi, O., Myers, J., and Nanni, A., "GFRP Reinforcements in Box Culvert Bridge: A Case Study After Two Decades of Service," *Concrete Pipe and Box Culverts, ASTM STP1601*, J. J. Meyer and J. Beakley, Eds., ASTM International, West Conshohocken, PA, 2017, pp. 75–88, http://dx.doi.org/10.1520/STP160120160119[4]

ABSTRACT

Corrosion-resistant glass fiber reinforced polymer (GFRP) composite bars are emerging as an alternative for traditional steel reinforcement in concrete structures exposed to aggressive environments such as bridges and box culverts. Although GFRP eliminates the problems related to corrosion of steel reinforcement, its long-term behavior in commercial applications needs to be confirmed. A box culvert bridge consisting of precast concrete units entirely reinforced with GFRP bars (constructed in 1999, on Walker Avenue in the city of Rolla, Missouri) was chosen as a case study. It replaced the original bridge that was built in the early 1980s and had been diagnosed as unsafe to operate due to excessive corrosion of encased steel pipes. Material samples were extracted from different locations of the box culvert and analyzed to monitor possible changes in GFRP and concrete after more than 16 years of service. Initially, carbonation depth, pH, and chloride diffusion measurements were performed on concrete cores surrounding the GFRP bars. Subsequently, scanning electron microscopy (SEM) imaging and energy dispersive X-ray spectroscopy (EDS) were conducted on GFRP samples to monitor any microstructural degradation or change in

Manuscript received August 30, 2016; accepted for publication January 3, 2017.

[1]University of Miami, College of Engineering, 1251 Memorial Dr., Coral Gables, FL 33146 http://orcid.org/0000-0003-0863-6014

[2]Missouri University of Science and Technology, Dept. of Civil, Architectural and Environmental Engineering, 1401 N. Pine St., Rolla, MO 65409 http://orcid.org/0000-0001-5269-8218

[3]University of Miami, College of Engineering, 1251 Memorial Dr., Coral Gables, FL 33146 http://orcid.org/0000-0003-2678-9268

[4]ASTM Symposium on *Concrete Pipe and Box Culverts* on December 7, 2016 in Orlando, Florida.

chemical compositions. In addition, glass transition temperature (T_g) of the resin and fiber content were determined and results were compared with pristine samples produced in 2015. Results from the concrete tests were consistent with expected values corresponding to the type and age of the structure. The SEM images and EDS test did not show any signs of GFRP microstructural deterioration. Moreover, T_g and fiber content of GFRP coupons were comparable to values from samples tested in 2015. The results of this study validate the notation that GFPR material properties are maintained during two decades of service. Hence, using GFRP internal reinforcement in box culverts eliminates corrosion problems, reduces long-term maintenance costs, and increases the service life of a structure.

Keywords

glass fiber reinforced polymer (GFRP), reinforced concrete, box culvert, scanning electron microscopy (SEM), dispersive X-ray spectroscopy (EDS)

Introduction

Precast reinforced concrete (RC) box culverts are rather common solutions for stream or open drain flow under roads or railways and are constructed as single or multiple structures. Precast RC box culverts offer advantages such as enhanced quality control, lower cost (due to mass production), and shorter installation time all of which led to a boom in their use in the 1970s [1]. Thus, most culverts are now more than 30 years old, and the ones made of corrugated steel or steel-RC show high levels of deterioration. The exposure of box culverts to a combination of aggressive environments such as water and chlorides from deicing salts accelerates the corrosion and leads to loss of serviceability.

The use of glass fiber reinforced polymer (GFRP) bars as flexural and shear reinforcement of concrete members is rapidly increasing especially due to the corrosion resistance properties of these composite materials [2]. GFRP bars were implemented in the Walker Avenue box culvert bridge in Rolla, MO, as an alternative to internal steel reinforcement in order to extend the service life beyond that of conventional steel-RC construction. However, confirmation of GFRP long-term durability is still necessary for the widespread acceptance of this technology in field applications. Accelerated laboratory tests are used to investigate GFRP durability in concrete structures by exposure to simulated concrete pore water solution at high temperatures. These tests are typically performed in an alkaline environment that is different from that present in field structures [3]. Conversely, monitoring the performance of existing RC structures would give a real indication of GFRP durability and, due to the inherent difficulty, only a few studies of this type are available [4–6]. In order to contribute to the existing body of technical literature and especially to investigate the effectiveness of GFRP bars in thin-walled RC structures, this study is intended to characterize GFRP bars and surrounding concrete after 16 years of service in a box culvert.

First, pH, carbonation depth, and chloride diffusion measurements were conducted on concrete cores to learn about the concrete environment. Next, microscopic examination and tests including scanning electron microscopy (SEM), energy dispersive X-ray spectroscopy (EDS), glass transition temperature (T_g), and fiber content were performed on GFRP coupons to determine possible changes in microstructure properties. Because no test results from GFRP bars at the time of construction were available, findings were compared with results of similar tests performed on the bar produced in 2015 by the same manufacturer, which served as a benchmark. In 1999, the GFRP bars used for these precast RC culverts were made of the E-glass fiber, while E-CR glass fiber was used in GFRP bars produced in 2015. The matrix in 1999 was polyester, while vinyl ester resin is used in the current production. Higher amounts of catalyst and promoters were also used in the bent bars compared to the straight bars due to the differences in curing procedures [7]. **Table 1** provides the fiber type, resin formulation, and the detailed catalysts/additives used in the production of GFRP bars. The most important implication of this change in constituents over time (common among all North American pultruders) is that the durability of the resulting GFRP bars is greatly enhanced given the stability of both E-CR glass and vinyl ester compared to E-glass and polyester.

Walker Box Culvert Bridge

The Walker box culvert bridge was constructed in 1999 on Walker Avenue in the city of Rolla, Missouri, to replace the original bridge, which was made of three concrete-encased corrugated steel pipes (with a diameter of 1.07 m) and had become unsafe to operate due to excessive corrosion of the steel pipes. The new bridge is 10.97 m (36 ft.) wide, consisting of 18 boxes of 1.5 by 1.5 m (4.92 by 4.92 ft.) and the thickness of 150 mm (5.90 in.). The precast RC boxes were arranged in two rows of nine and designed in accordance with American Association of State Highway and Transportation Officials' (AASHTO) design guidelines [8]. The old and new bridges are shown in **Fig. 1**. The RC boxes were entirely reinforced with No. 2

TABLE 1 Constituents of GFRP bars produced in 1999 and 2015 [7].

Year of Production	Fiber	Type	Resin		Additive and Filler	Catalyst
			Type	Formulation		
1999	E-glass	Bent bars	Polyester	Aropol 7420 (73 %)	Styrene (5 %), ASP 400 (21 %) and DMA	MEKP and Cobalt
2015	E-CR glass	Straight bars	Vinyl Ester	VEX 10-962 CoRezyn	Styrene & ASP 400	BPO

Note: ASP = aluminum silicate pigments; DMA = dimethylamine; MEKP = methyl ethyl ketone peroxide; BPO = benzoyl peroxide.

FIG. 1 Old (left) and new (right) Walker Ave. bridge.

GFRP bars (nominal diameter of 6.3 mm) pre-bent and cut to size by the manufacturer. The guaranteed properties of the bars used in the original construction are provided in **Table 2**.

Following AASHTO recommendations [8], circumferential reinforcement was provided in the top panel at every 100 mm (3.94 in.). Circumferential reinforcement at every 200 mm (7.87 in.) was also provided in the walls and bottom panel to facilitate the reinforcement cage construction. The bars were tied together using plastic ties. The 25.4-mm (1 in.) concrete cover was maintained using plastic wheel spacers. Box culvert details such as layout of the bars and concrete cover can be seen in **Fig. 2** [1]. Additionally, **Fig. 2** shows a completed GFRP cage before concrete casting. Conventional concrete made of portland cement, fly ash, tap water, and Missouri River aggregate with a maximum aggregate size of 9.5 mm (3/8 in.) was used to produce the units. The boxes were cast using a dry cast process that uses low frequency, high amplitude vibration to distribute the mix. The measured concrete compressive strength was 42.7 MPa (6.2 ksi) [1].

Sample Extraction and Preparation

Technically competent personnel performed the extraction of concrete cores from the culverts in 2015 after 16 years of service from the time of construction. Four 101.6-mm (4 in.) diameter concrete cores were extracted from the bottom of two culverts (**Fig. 3**). GFRP coupons were extracted from the cores as shown in **Fig. 4**,

TABLE 2 Guaranteed properties of the GFRP bars used in Walker Ave. bridge.

Reinforcement	Diameter (mm)	Tensile Strength f^* (MPa)	Elastic Modulus E_f (GPa)	Rapture Strain (%)
No. 2 GFRP	6.3	757.9	40.7	1.9

Note: 1 mm = 0.0394 in.; 1 MPa = 0.145 ksi.

FIG. 2 Layout of the box culvert (left) and GFRP cage before concrete casting (right).

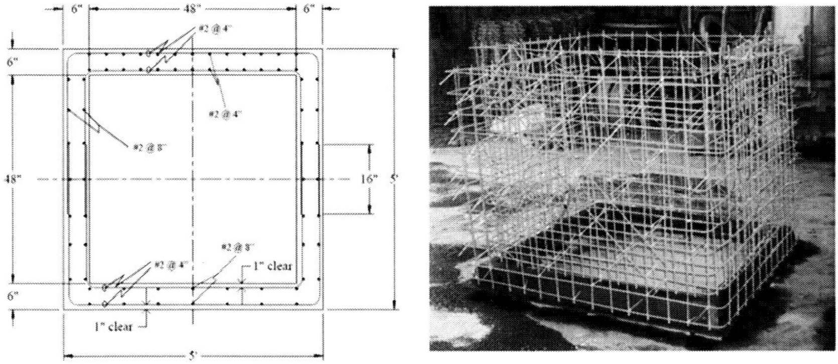

and two of them were sliced to an approximate width of 10 mm (0.4 in.) using a diamond saw for microscopic examination.

The surface of the GFRP slices was prepared by sanding using different levels of sandpaper (i.e., Nos. 180, 300, 600, and 1200) and utilizing dedicated grinding and polishing equipment. Fine polishing completed the specimen preparation using a wet-polishing agent and polycrystalline diamond paste. Prior to imaging, specimens were placed in an oven at 60°C (140°F) for 24 h to remove moisture produced during polishing. Additionally, a concrete sample was prepared following the same procedure to monitor the concrete-GFRP interface. Because GFRP and concrete are nonconductive materials, an ion sputtering device was used to coat the samples with gold prior to SEM examination as shown in **Fig. 5**. The specimens used in SEM

FIG. 3 Sample extraction from box culvert units.

FIG. 4 GFRP coupons extracted from the concrete cores.

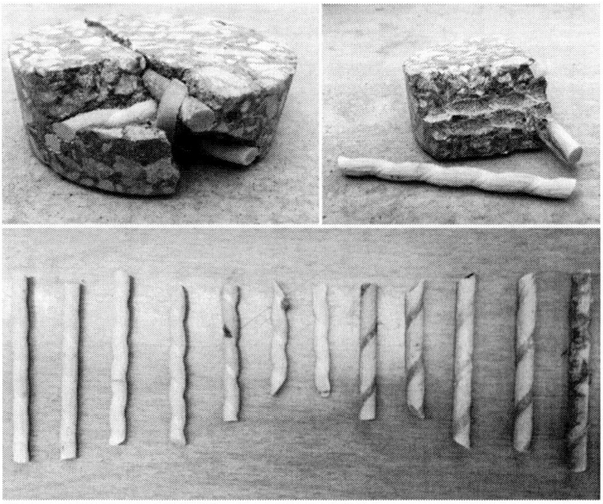

imaging were also utilized in EDS analysis. Additionally, the coupons were cut in appropriate sizes for T_g and fiber content measurements.

Concrete Characterization

pH MEASUREMENT

The pH measurement was performed to provide a qualitative estimate of concrete alkalinity. The pH measurement approach proposed by Grubb and coworkers [9] was followed because it provides a more precise assessment compared to the

FIG. 5 Prepared concrete and GFRP samples for SEM imaging.

ASTM F710-11, *Standard Practice for Preparing Concrete Floors to Receive Resilient Flooring* [10], method, which typically underestimates the pH due to excessive wetting of the concrete surface. First, the concrete surface at the depth of 24.5 to 50.8 mm (1 to 2 in.) of the cores was ground using sandpaper and diluted in distilled water with a 1:1 ratio. Then, the pH strip was used to evaluate the alkalinity of the solution (**Fig. 6**); pH values between 11 and 12 were measured, which meet expectation for that type of concrete and age [9]. The procedure was performed in three different locations of the core and consistent results were obtained.

CARBONATION DEPTH

As a result of carbonation, the initial high pH value of the cement paste may drop to values below nine, forming a low pH layer of concrete at the surface. A carbonation depth equal to the concrete cover may be responsible for steel corrosion

FIG. 6 Concrete pH measurement.

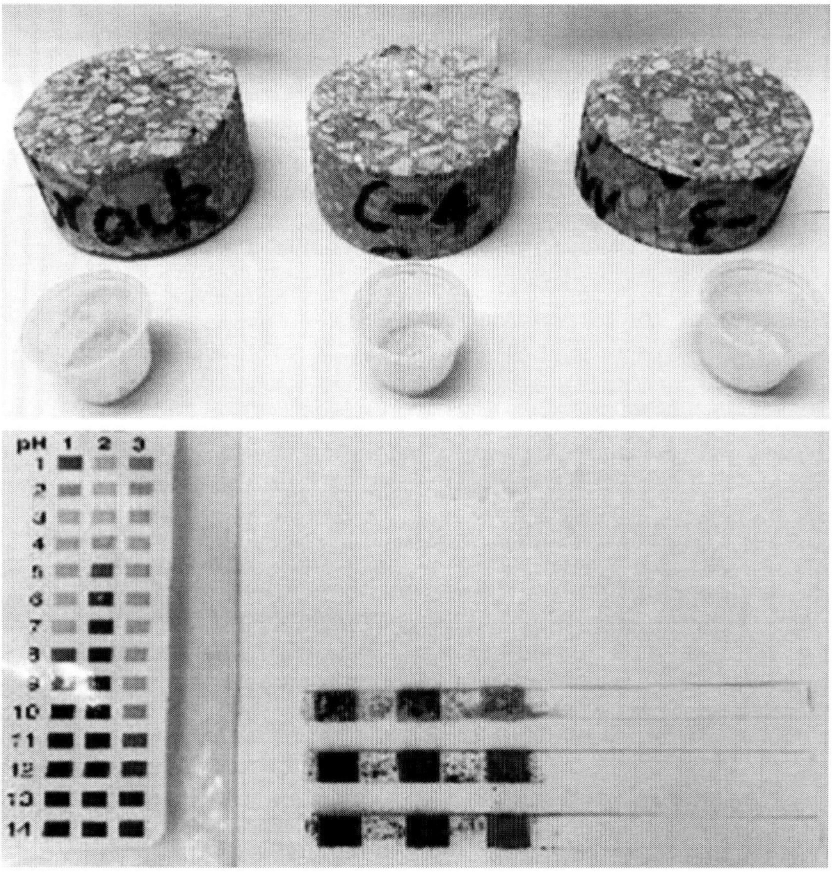

FIG. 7 Carbonation depth measurement.

initiation. The carbonation depth was measured by spraying the 1 % solution of phenolphthalein in 70 % ethyl alcohol on freshly fractured concrete surfaces [11]. The colorless solution turned to pink/purple when the pH was higher than nine and stayed colorless otherwise. **Fig. 7** shows extracted concrete samples where the concrete surface was chipped off at the edge and sprayed with the phenolphthalein solution. No indication of concrete carbonation was observed using this method.

Although no carbonation of concrete can be considered beneficial to steel rebars because the pH remains at high values, the opposite is true for GFRP reinforcement, which is more sensitive to high alkalinity. Thus, the GFRP bars extracted from these cores were subject to an aggressive environment over the 16-year service life.

CHLORIDE DIFFUSION MEASUREMENT

An adaptation of the rapid migration test using silver nitrate solution was used to determine the chloride diffusion in the concrete samples. Two concrete samples were cut in order to provide fresh split surfaces. A 0.1-mol/L silver nitrate solution was poured on the entire cut surface [12]. In the presence of chloride, a clearly visible white/silver precipitation takes place on the surface, while in the absence of chlorides, the solution reacts with the hydroxides present in the concrete, changing the surface color to brown. No clear evidence of chloride diffusion was observed in all the tested specimens using this method. It was noticed that the surface became darker, to a color similar to brown, although there was no visible gray area (**Fig. 8**). This result was expected because there was a low possibility of chloride diffusion at the bottom of the culvert, although the top part was exposed to deicing salt.

GFRP Characterization

Because the possible degradation mechanisms of GFRP are a function of water ingress, a durability characterization of smallest bar size (i.e., No. 2) represents the most conservative case.

FIG. 8 Chloride diffusion measurement: Pristine sample (left) compared to the tested sample (right).

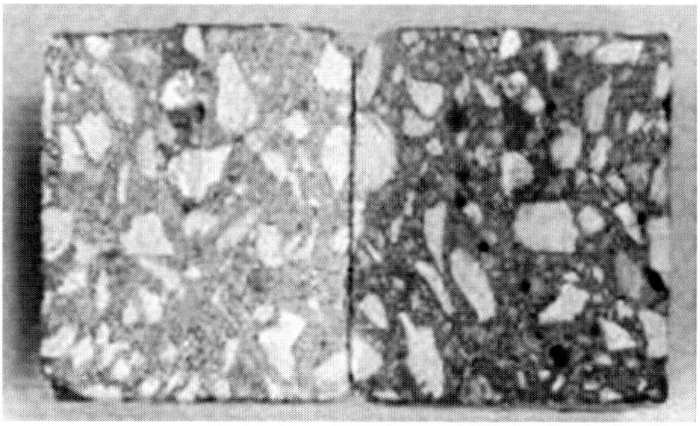

SEM IMAGING

The full cross sections of two slices of No. 2 GFRP bars were scanned at different levels of magnification and images were taken at random locations. Representative images are shown in **Fig. 9**. Attention was paid to the areas in the vicinity of the bar edges because possible degradation due to chemical attack starts at the GFRP-concrete interface. SEM analysis confirmed that there was no sign of deterioration in the GFRP coupons. Glass fibers were intact without a loss of the cross-sectional area. Similarly, fibers were surrounded by the resin matrix and no sign of loss of the bond between matrix and fiber was observed. Additionally, the GFRP-to-concrete interfacial bond appeared to be maintained properly, and no sign of bond degradation nor loss of contact was observed as presented in **Fig. 10**.

FIG. 9 SEM images of GFRP bar after 16 years of service in magnification levels of 200× (left) and 800× (right).

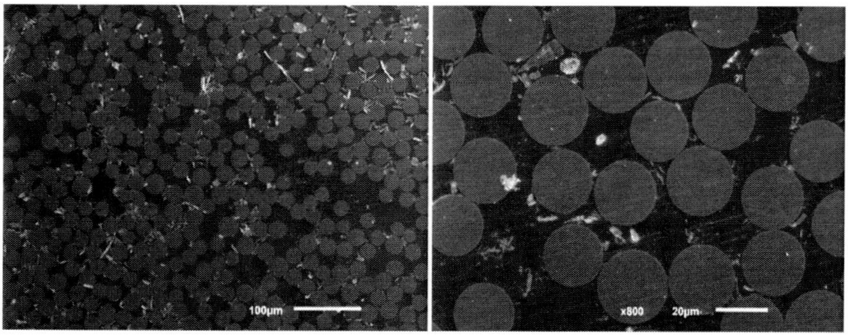

FIG. 10 Concrete to GFRP interface.

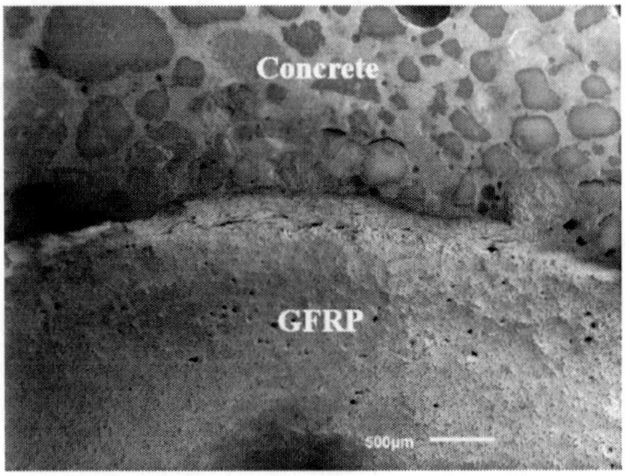

EDS ANALYSIS

EDS was performed at five selected locations of the No. 2 GFRP slices, with a focus on the edge of the bar, to identify existing chemical elements in GFRP bars. The results were compared with pristine samples produced in 2015 from the same manufacturer. The results are shown in **Fig. 11** and **Fig. 12**, where the vertical axis

FIG. 11 Results of the EDS analysis performed on GFRP samples after 16 years of service.

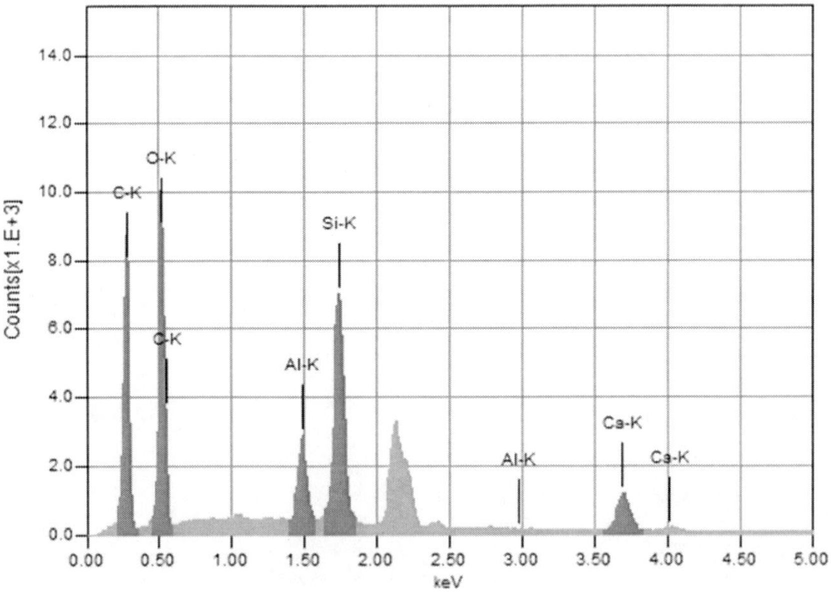

FIG. 12 Results of the EDS analysis performed on control GFRP samples produced in 2015.

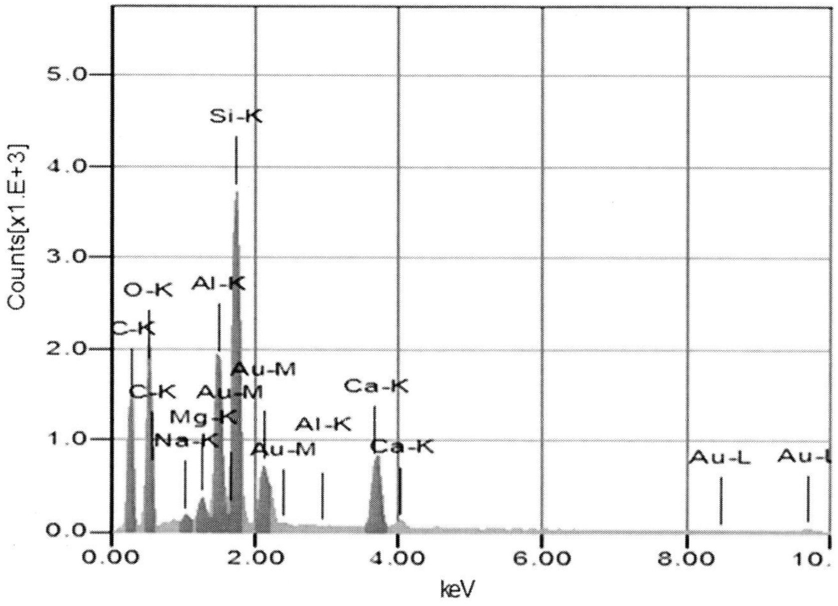

corresponds to the counts (number of X rays received and processed by the detector) and the horizontal axis presents the energy level of those counts.

Silcon (Si), aluminum (Al), calcium (Ca; from glass fibers) and carbon (C; from the matrix) were the predominant chemical elements in the extracted samples, which were also identical to the control samples. Although there is a variation in fiber/resin constituents for GFRP bars produced in 2015 compared to the ones manufactured in 1999 (**Table 1**), the only difference in detected elements between the two was the presence of magnesium (Mg) and sodium (Na) in the control samples, which was not found in the extracted ones. The difference in fiber and resin constituents may be the reason for this dissimilarity. Comparing the result of EDS analysis performed on the extracted and control samples confirmed that no change in chemical composition of fiber and matrix occurred after 16 years of service.

GLASS TRANSITION TEMPERATURE (T_g)

The changes in T_g of the polymer matrix were determined by a performing dynamic mechanical analysis (DMA) test on three specimens. A T_g higher than 100°C (212°F) is desirable as a critical parameter [13]. Rectangular specimens with dimensions of 1 by 5 by 50 mm (0.04 by 0.2 by 2.0 in.) were extracted from the bars according to ASTM E1640, *Standard Test Method for Assignment of the Glass Transition Temperature by Dynamic Mechanical Analysis* [14]. The DMA test was

TABLE 3 Results of T_g performed on extracted and pristine GFRP bars.

T_g^c			T_g^s			
No. of Samples	Average (°C)	CoV (%)	No. of Samples	Average (°C)	CoV (%)	Ratio (T_g^s/T_g^c)
3	81.0	16.9	3	111.9	2.5	1.31

Note: °F = 1.8°C + 32; CoV = coefficient of variation.

performed with a three-point-bending fixture for a temperature ranging from 30°C to 130°C (86°F to 266°F) and a heating rate of 1°C/min (1.8°F/min). Due to lack of T_g test data on GFRP bars at the time of construction, T_g tests were performed on samples from pristine bars produced in 2015 from the same manufacturer to serve as a benchmark for comparison. **Table 3** provides the result summary, where T_g^c and T_g^s refer to the glass transition temperature of the control and in-service GFRP samples, respectively.

The T_g of the extracted samples was 30% higher than the control samples pultruded in 2015. As described in **Table 1**, due to changes in glass fibers, resin formulation, additive, and catalysts of the bars manufactured in 2015 compared to the ones produced in 1999, a direct comparison is not possible. In general, T_g is expected to increase over time due to cross-linking of the resin if it is not 100 % cured at the time of manufacturing.

FIBER CONTENT

The fiber content of GFRP samples was determined following ASTM D2584, *Standard Test Method for Ignition Loss of Cured Reinforced Resins* [15]. Three samples were tested for change in mass. Samples were first placed inside the furnace for 40 min at 425°C (797°F) and then were left inside the furnace at 700°C (1292°F) for 30 min to completely burn off the resin. The weight of sand particles and wrapping strand at the GFRP surface was also eliminated to provide a precise estimation of fiber content. The result was compared with the same test performed in 2015. **Table 4** shows the summary of the result where α_c and α_s correspond to fiber ratio of control and extracted samples, respectively. The measured fiber content after 16 years of field exposure was consistent with the expected values and well above the minimum fiber content requirement of 70 % by mass [16].

TABLE 4 Results of fiber content measurement performed on extracted and GFRP samples produced in 2015.

α_c			α_s		
No. of Samples	Average (%)	CoV (%)	No. of Samples	Average (%)	CoV (%)
4	75.7	1.2	4	82.38	4.0

Note: CoV = coefficient of variation.

Conclusions

According to the results of the experimental tests performed on extracted concrete cores and GFRP rebars after 16 years of service as concrete reinforcement in a precast box culvert, the following observations can be made:

- The concrete pH was in the range of 11 to 12, which is consistent with the concrete type and age.
- No indication of carbonation and chloride diffusion was observed in the concrete samples.
- Microscopic examination did not show any GFRP degradation. Fibers did not lose any cross-sectional areas, the matrix was intact, and no damage was observed at the fiber-matrix interface. Additionally, the concrete-GFRP interface was maintained properly and no interfacial bond loss was observed.
- T_g of the extracted GFRP samples was 30 % higher than that of the control samples produced in 2015 by the same manufacturer.
- The result of fiber content measurement of extracted GFRP bars was consistent with the pristine bars performed in 2015 confirming that there was no apparent loss of fiber content in GFRP bars.

This study confirms that GFRP bars maintained their microstructural integrity after 16 years of service in a GFRP box culvert bridge. The results of this paper present additional evidence that GFRP bars can eliminate the corrosion problem of black steel rebar in box culvert structures and increase their service life.

ACKNOWLEDGMENTS

The authors gratefully acknowledge: (a) the University Transportation Center "Research on Concrete; Applications for Sustainable Transportation (RE-CAST)" under grant US DOT, DTRT13-G-UTC45; (b) the National Science Foundation (NSF) and its industrial members for the support provided to the Industry/University Center for Integration of Composites into Infrastructure under grant NSF IIP-1439543; and (c) Hughes Brothers, Inc., for their openness in describing the past and present production processes and resin formulations.

Findings and opinions expressed herein, however, are those of the authors alone and do not necessarily reflect the views of the sponsors.

References

[1] Alkhrdaji, T. and Nanni, A., *Construction and Long-Term Monitoring of a Concrete Box Culvert Bridge Reinforced with GFRP Bars*, Technical Report: #RDT01-016, Center for Infrastructure Engineering Studies, Rolla, MO, 2001.

[2] Nanni, A., De Luca, A., and Jawaheri Zadeh, H., *Reinforced Concrete with FRP Bars: Mechanics and Design*, CRC Press, Boca Raton, FL, 2014.

[3] Micelli, F. and Nanni, A., "Durability of FRP Rods for Concrete Structures," *Constr. and Build. Mater.*, Vol. 18, No. 7, 2004, pp. 491–503.

[4] Mukherjee, S. and Arwikar, S. J., "Performance of Glass Fiber-Reinforced Polymer Reinforcing Bars in Tropical Environments—Part II: Microstructural Test," *ACI Struct. J.*, Vol. 102, No. 6, 2005, pp. 816–822.

[5] Mufti, A., Banthia, N., Benmokrane, B., Boulfiza, M., and Newhook, J., "Durability of GFRP Composite Rods," *Concr. Int.*, Vol. 29, No. 2, 2007, pp. 37–42.

[6] Gooranorimi, O., Bradberry, T., Dauer, E., Myers, J., and Nanni, A., *FRP Reinforcement for Concrete: Performance Assessment and New Construction, Volume I: Sierrita De La Cruz Creek Bridge*, Technical Report: #00042134-04-103A, University of Miami, Coral Gables, FL, 2016.

[7] Gremel, D., Private communication, Hughes Brothers, Seward, NE, 2016.

[8] AASHTO, *Standard Specifications for Highway Bridges*, 16th ed., American Association of State Highway and Transportation Officials, Washington, DC, 1996.

[9] Grubb, J., Limaye, H., and Kakade, A., "Testing pH of Concrete: Need for A Standard Procedure," *Concr. Int.*, Vol. 29, No. 4, 2007, pp. 78–83.

[10] ASTM F710-11, *Standard Practice for Preparing Concrete Floors to Receive Resilient Flooring*, ASTM International, West Conshohocken, PA, 2008, www.astm.org

[11] RILEM Technical Committees, "CPC-18 Measurement of Hardened Concrete Carbonation Depth," *Mater. Struct.*, Vol. 21, No. 126, 1988, pp. 453–455.

[12] Otsuki, N., Nagataki, S., and Nakashita, K., "Evaluation of $AgNO_3$ Solution Spray Method for Measurement of Chloride Penetration into Hardened Cementitious Matrix Materials," *ACI Mater. J.*, Vol. 89, No. 4, 1992, pp. 587–592.

[13] ACI 440.6-08, *Specification for Carbon and Glass Fiber-Reinforced Polymer Bar Materials for Concrete Reinforcement*, American Concrete Institute, Farmington Hills, MI, 2008.

[14] ASTM E1640, *Standard Test Method for Assignment of the Glass Transition Temperature by Dynamic Mechanical Analysis*, ASTM International, West Conshohocken, PA, 2013, www.astm.org

[15] ASTM D2584, *Standard Test Method for Ignition Loss of Cured Reinforced Resins*, ASTM International, West Conshohocken, PA, 2011, www.astm.org

[16] AC-454, *Acceptance Criteria for Fiber-Reinforced Polymers (FRP) Bars for Internal Reinforcement of Concrete Members*, International Code Council–Evaluation Service, Whittier, CA, 2015.

STP 1601, 2017 / available online at www.astm.org / doi: 10.1520/STP160120160109

Corey L. Haeder[1]

Structural Design of ASTM C361 Low Head Pressure Pipe Joints

Citation

Haeder, C. L., "Structural Design of ASTM C361 Low Head Pressure Pipe Joints," *Concrete Pipe and Box Culverts, ASTM STP1601*, J. J. Meyer and J. Beakley, Eds., ASTM International, West Conshohocken, PA, 2017, pp. 89–117, http://dx.doi.org/10.1520/STP160120160109[2]

ABSTRACT

The current published version of ASTM C361, *Standard Specification for Reinforced Concrete Low-Head Pressure Pipe*, does not provide guidance for structural design of the concrete pipe joint. With the adoption of ASTM C361 by the U.S. Bureau of Reclamation, some of the information that was formerly included in their low-head pressure pipe publication (USBR M-1, *Standard Specifications for Reinforced Concrete Pressure Pipe*) is no longer in print. Discussion of the factors involved in the design of concrete joints for low-head pressure pipe are presented as well as a recommended design procedure and comparison with assumptions included in the USBR M-1 publication.

Keywords

concrete, pipe, pressure, head, reinforced, joints

Introduction

The purpose of this paper is to discuss structural design of reinforced concrete low-head pressure pipe (RCLHPP) joints manufactured with concrete bells and spigots. This discussion will include a review of existing design methods, comparison of these methods with three-dimensional finite element analysis (3D FEA) of the pipe sections, and development of a rational method to determine reinforcement requirements for concrete bell sections. The author acknowledges that additional issues affect joints, including—but not

Manuscript received August 17, 2016; accepted for publication November 9, 2016.
[1]Engineering Dept., Forterra Building Products, 6655 Wedgwood Rd., Maple Grove, MN 55311 http://orcid.org/0000-0002-2080-9734
[2]ASTM Symposium on *Concrete Pipe and Box Culverts* on December 7, 2016 in Orlando, Florida.

limited to—joint shear due to nonuniform longitudinal support. For simplicity of analysis and presentation of the information included in this paper, only the stresses caused by internal pressure, vertical earth loads, and gasket compression have been included.

The following publications are available for specification of RCLHPP:

- USBR M-1, *Standard Specifications for Reinforced Concrete Pressure Pipe*, U.S. Bureau of Reclamation, November 1, 1991 [1].
  - Material and design guidelines as well as minimum reinforcement areas for pipe wall sections and concrete bells for various internal pressures and fill height classes.
  - Joint details for concrete bell and spigot, steel bell with concrete spigot, and steel bell and steel spigot joints.
  - Production tolerances and testing requirements.
  - Standard retired in 2011 and replaced with ASTM C361, *Standard Specification for Reinforced Concrete Low-Head Pressure Pipe*.
  - Standard joint details (R1, R2, R3, and R4) incorporated into projects as a standard plate.
- ASTM C361, *Standard Specification for Reinforced Concrete Low-Head Pressure Pipe*, ASTM International, 2014a [2].
  - Material and design guidelines as well as minimum reinforcement areas for pipe walls only.
  - Production tolerances and testing requirements.
  - Lacks joint details or joint structural design information.
  - Updated rational design incorporated in 2011 edition adding crack control and shear/radial tension checks to the design process.
- AWWA C302, *Reinforced Concrete Pressure Pipe, Noncylinder Type*, American Water Works Association, 2011 [3].
  - Generic material specification that does not cover design of wall sections or details of joints.
- AWWA M9, *Concrete Pressure Pipe*, third ed., American Water Works Association, 2008 [4].
  - Specification providing design guidance similar to ASTM C361 but does not include joint structural design guidance.

Both the USBR M-1 standard and ASTM C361 follow a general classing system including a letter (A, B, C, or D) indicating the fill height in 5-ft increments (A = 5 ft, D = 20 ft) followed by a numeric pressure class (head in ft) in 25-ft increments. Examples:

- 36-in. A25 = 36-in. diameter, 5-ft maximum fill, 25-ft maximum design pressure
- 72-in. D100 = 72-in. diameter, 20-ft maximum fill, 100-ft maximum design pressure

With the retirement of the USBR M-1 standard, the USBR has been including joint designs directly in project plans. This covers the needs of USBR projects, but projects specified according to ASTM C361 by other entities do not have the ability to directly reference these standards and would benefit from a standard joint design process.

Bureau of Reclamation Joint Design Procedures

As detailed in the USBR M-1 standard, all RCLHPP joints are to be one of four joint types. Each type incorporates a rubber gasket of circular cross section as the only means of sealing the joint in a fully confined installation with containment provided on four sides of the gasket. Descriptions of the four joint types are as follows:

- R1—Concrete spigot with confining groove and steel bell band with concrete encasement
- R2—Steel bell and spigot rings cast integral with the pipe sections
- R3—Concrete bell with confining shoulder and concrete spigot with confining shoulder
- R4—Concrete bell and concrete spigot with confining groove

The most current version of the USBR M-1 standard (1991) assumes that rubber gaskets impart 200 lbf/in. of contact with the bell surface. This is an arbitrary assumption that is intended to provide an upper bound to the potential gasket forces involved in a concrete pipe joint. Prior to changing to 200 lbf/in., the standard assumed 150 lbf/in. The reason for the change was to address field performance issues related to failures in the concrete bell by increasing its required strength. Following are the variables that impact the resulting gasket forces:

- Rubber material stiffness
- Rubber gasket diameter in the installed condition, considering effects of gasket stretch
- Joint geometry
- Production tolerances that create maximum or minimum gasket compression
- Differential loading across a joint due to backfill and bedding soils

Each of these variables impacts the force a gasket imparts on the bell and varies from joint to joint and around the perimeter of a given joint. Rationalizing the impact of these variables into a highly refined design process would be extremely difficult and given the importance of joint performance, conservative assumptions, such as 200 lbf/in., are prudent. The added expense associated with specification and field control of a more refined design is easily offset by the potential cost of repairing failures.

Gasket manufacturers are able to develop tooling that can be used to provide force/deformation curves for a given pipe manufacturer's joint details and specified gasket size and material. Consulting with the manufacturer during the design phase can help to establish if the 200 lbf/in. design assumption is adequate or if a lower/higher number is more appropriate. This type of discussion should be addressed with the gasket supplier again any time the joint forming equipment or manufacturing processes change.

Based on the assumed gasket force, the following simplified procedure is specified by the USBR M-1 standard:

36-in. Inside Diameter Reinforced Concrete Pipe (RCP) Example

Step 1: Determine the force that the gasket imparts onto the bell (HEAD = 100 ft).

Inside diameter (ID) of the bell in inches	ID = 43 in. (given by manufacturer)
Bell force from gasket (BF1)	BF1 = 0.5 * (ID * 200 lbf/in.)
	BF1 = 4,300 lb
Internal pressure on 1.75-in. joint length	BF2 = HEAD * 62.4/144 * 1.75 * ID * 0.5
	BF2 = 1,630 lb
Total bell force	BF = BF1 + BF2
	BF = 5,930 lb

Step 2: Determine the steel stress that is allowed (strain based).

Allowable steel stress (STRESS)	STRESS = 17,000 psi − 35 * HEAD
	STRESS = 13,500 psi

Step 3: Determine the steel area required (force divided by stress).

Area of steel required in the bell (ASREQ)	ASREQ = BF/STRESS
	ASREQ = 0.44 sq. in.

Step 4: Provide details for bell reinforcement.

Equal distribution over a length = 1.75 × joint length per USBR M-1 procedures

The USBR M-1 standard provides a table for bell reinforcement for all sizes and classes of pipe available in the standard based on the information illustrated previously. Bell reinforcement is to be applied as detailed on the R-3 or R-4 joint standards as shown in **Fig. 1** and **Fig. 2** where the minimum bell reinforcement is to meet the requirements of the minimum bell reinforcement table shown in **Table 1**. Minimum pipe wall reinforcement is depicted in **Table 2**, similar to the excerpt shown.

As shown in **Table 1**, the bell reinforcement varies with the pressure class of a given design, but it does not vary with the fill class of the design. Because the bell is integrally a part of the pipe section, the stresses induced by fill on the pipe wall will carry into the bell. Consequently, the bell reinforcement should vary with fill height as well as pressure class. Currently, the impact of bell stresses associated with fill height is not part of the bell reinforcement design process. However, given the arbitrary 200 lbf/in. gasket force requirement, the soil load stresses may be accounted for indirectly by the design process. The relative impact fill height has on bell stress is discussed in the following sections.

Finite Element Analysis

In order to determine the impact fill height has on bell stresses, a 3D FEA of pipe sections was performed. The analysis was performed using Simulation Mechanical from Autodesk with the following FEA model assumptions:

- Autodesk Simulation Mechanical.
- Half pipe section—8-ft full length.
- Symmetrical boundary condition.

FIG. 1 USBR M-1 joint detail for R-3 joints.

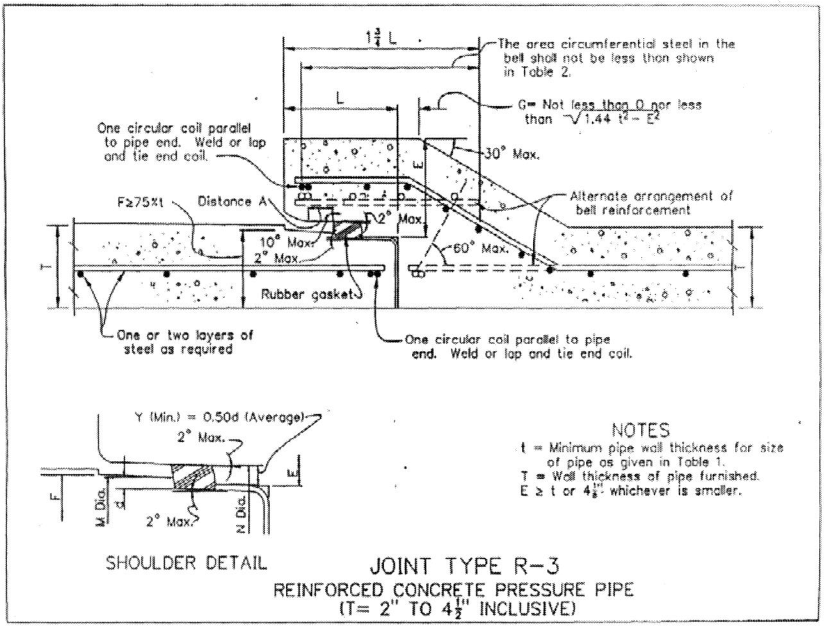

FIG. 2 USBR M-1 joint detail for R-4 joints.

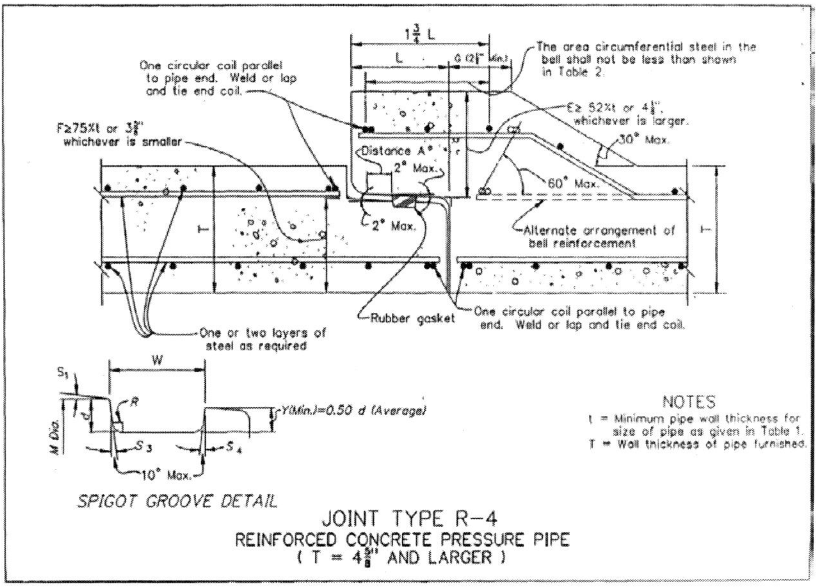

TABLE 1 USBR M-1 minimum bell reinforcement (portion).

	Minimum Bell Reinforcement in Square Inches to be Distributed in 1-3/4 L																										
Internal diameter of pipe in inches	12	15	18	21	24	27	30	33	36	39	42	45	48	51	54	57	60	63	66	69	72	78	84	90	96	102	108
Class																											
A-25																											
B-25																											
C-25	0.11	0.13	0.15	0.17	0.19	0.23	0.25	0.27	0.29	0.31	0.33	0.35	0.38	0.40	0.42	0.44	0.46	0.48	0.50	0.52	0.54	0.58	0.62	0.66	0.70	0.74	0.78
D-25																											
A-50																											
B-50																											
C-50	0.13	0.15	0.18	0.20	0.22	0.26	0.29	0.31	0.34	0.36	0.38	0.41	0.43	0.45	0.48	0.50	0.52	0.55	0.57	0.59	0.62	0.66	0.71	0.76	0.81	0.85	0.90
D-50																											
A-75																											
B-75																											
C-75	0.15	0.17	0.20	0.23	0.25	0.30	0.33	0.36	0.39	0.41	0.44	0.47	0.49	0.52	0.55	0.57	0.60	0.63	0.65	0.68	0.71	0.76	0.82	0.87	0.92	0.98	1.03
D-75																											
A-100																											
B-100																											
C-100	0.17	0.20	0.23	0.26	0.29	0.34	0.38	0.41	0.44	0.47	0.50	0.53	0.56	0.60	0.63	0.66	0.69	0.72	0.75	0.78	0.81	0.87	0.93	0.99	1.05	1.12	1.18
D-100																											
A-125																											
B-125																											
C-125	0.19	0.23	0.26	0.30	0.33	0.39	0.43	0.47	0.50	0.54	0.57	0.61	0.65	0.68	0.72	0.75	0.79	0.82	0.86	0.89	0.93	1.00	1.07	1.14	1.21	1.28	1.35
D-125																											
A-150																											
B-150																											
C-150	0.22	0.26	0.30	0.34	0.38	0.45	0.50	0.54	0.58	0.62	0.66	0.70	0.74	0.78	0.82	0.86	0.90	0.94	0.98	1.02	1.06	1.14	1.22	1.30	1.38	1.46	1.54
D-150																											

TABLE 2 Excerpt of USBR M-1 minimum circumferential reinforcement for pipe wall.

	Minimum Circumferential Reinforcement in Square Inches per Linear Foot of Pipe																	
Internal Designated Diameter, in.	**36**									**72**								
Type of Reinforcement	**Circular**							**Elliptical**		**Circular**							**Elliptical**	
Wall Thickness, in.	**3.12**	**3.25**		**4.00**		**5.00**		**3.12**	**4.00**	**6.00**		**7.00**		**7.75**		**6.00**	**7.00**
Layers of Reinforcement	**Single**	**Inner**	**Outer**	**Inner**	**Outer**	**Inner**	**Outer**	**Single**	**Single**	**Inner**	**Outer**	**Inner**	**Outer**	**Inner**	**Outer**	**Single**	**Single**
Class																	
A-25	0.37	0.26	0.16	0.21	0.13	0.19	0.11	0.27	0.24	0.54	0.32	0.52	0.30	0.48	0.27	0.54	0.52
B-25	0.66	0.43	0.24	0.33	0.17	0.28	0.14	0.45	0.33	0.80	0.45	0.74	0.39	0.67	0.35	0.80	0.74
C-25		0.60	0.32	0.44	0.22	0.37	0.16	0.64	0.44	1.16	0.62	1.03	0.53	0.91	0.45	1.16	1.03
D-25		0.79	0.41	0.56	0.27	0.46	0.19		0.56	1.56	0.82	1.37	0.68	1.19	0.57		1.37
A-50	0.47	0.33	0.24	0.28	0.20	0.26	0.18	0.50	0.50	0.67	0.45	0.65	0.43	0.61	0.40	0.99	0.99
B-50	0.77	0.50	0.31	0.40	0.24	0.35	0.21	0.52	0.50	0.93	0.57	0.87	0.53	0.80	0.48	0.99	0.99
C-50		0.66	0.39	0.51	0.29	0.43	0.23	0.71	0.51	1.27	0.74	1.15	0.65	1.03	0.58	1.35	1.15
D-50		0.86	0.48	0.63	0.34	0.52	0.26		0.63	1.66	0.93	1.48	0.80	1.31	0.69		1.48

- Static stress with linear material model.
- Brick elements/isotropic material.
- Midside nodes included.
- Fairly high-strength concrete (4.5 million psi modulus of elasticity).
- Solid mesh.
- Gasket pressure = 200 lbf/in.
- Internal pressure = constant on inner perimeter (10.83 psi at 25-ft head, 43.32 psi at 100-ft head).
- Soil load at 5-ft = 4.167 psi vertical traction applied to entire pipe width.
- Soil load at 20-ft = 16.668 psi vertical traction applied to entire pipe width.
- Bearing reaction at 5-ft = 5.89 psi vertical traction applied over 90° bedding.
- Bearing reaction at 20-ft = 23.56 psi vertical traction applied over 90° bedding.
- 3D springs at ends of bell and spigot to maintain model stability (0.1 lbf/in. of deflection).
- Soil loads and bearing reaction not applied to extended bell if present.

To simplify the model, the effects of gravity on the fluid in the pipe as well as on the pipe wall were ignored.

As shown in **Figs. 3** and **4**, two sizes of pipe were chosen:

- 36-in. RCP—Standard R-4 joint with an extended bell
- 72-in. RCP—Standard R-4 joint without an extended bell

The analysis includes a 36-in. inside diameter pipe with extended bell to illustrate the impact the modified section properties of the bell section has on the overall pipe stresses both in the wall section and in the bell. The 72-in. inside diameter size

FIG. 3 FEA setup—36-in. pipe.

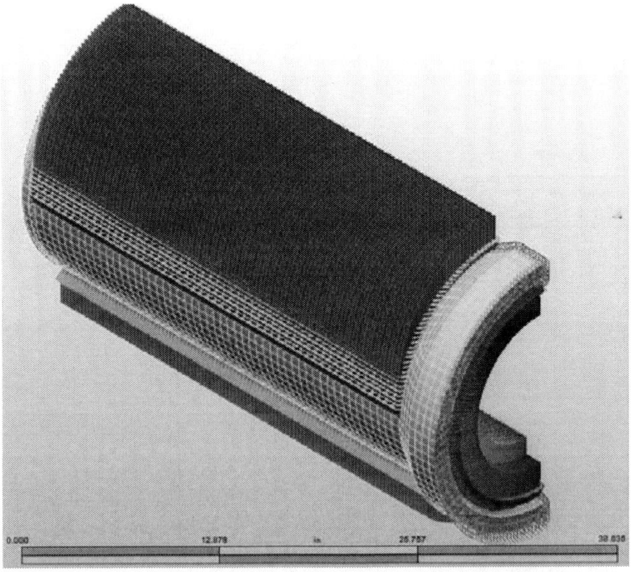

FIG. 4 FEA setup—72-in. pipe.

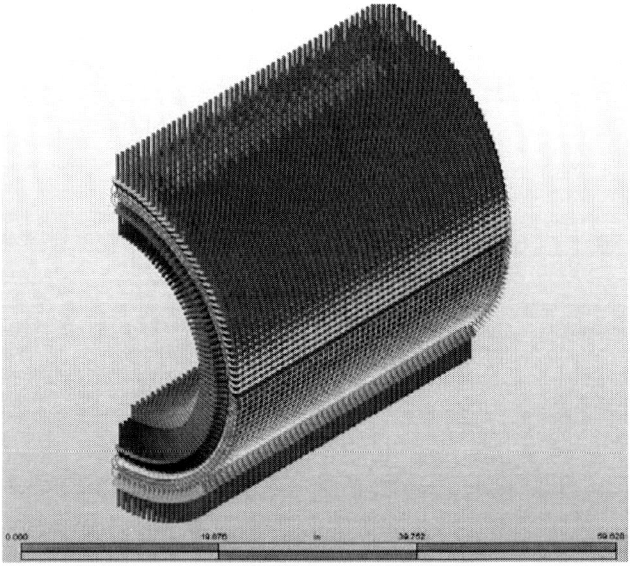

was chosen to illustrate how the stresses in a straight wall and bell differ from the extended bell. For each size of pipe, the following classes were modeled, both with and without a gasket, for a total of 16 models developed for the FEA.

- 36A25 and 36A100—with and without gasket
- 72A25 and 72A100—with and without gasket
- 36D25 and 36D100—with and without gasket
- 72D25 and 72D100—with and without gasket

Results of the FEA model simulations are presented in Figs. 5 through 20. The results shown are maximum principal stresses in psi (positive = tension) with the maximum bell stress shown in parentheses after the figure title. The figure on the left is the resulting stress plot with a gasket included. The figure on the right is the same without a gasket.

Figs. 5 through 12 for 36-in. pipe with an extended bell show a tendency for the bell section to act as a stiffening element or edge beam for the pipe. This can be attributed to the effective increased depth of the section at the bell providing a stiffer section and more effective use of reinforcement as the larger effective depth. The stresses in the bell are less than in the pipe wall with or without a gasket. The influence of an extended bell should be included in the design process.

Figs. 13 through 20 illustrate the results of the 72-in. RCLHPP models without an extended bell. A different trend is noticed in these cases.

Unlike the extended bell results for 36-in. diameter models, the 72-in. diameter models show that the stresses in the bell are somewhat less without the gasket but

FIG. 5 Results for 36A25 with gasket (251 psi).

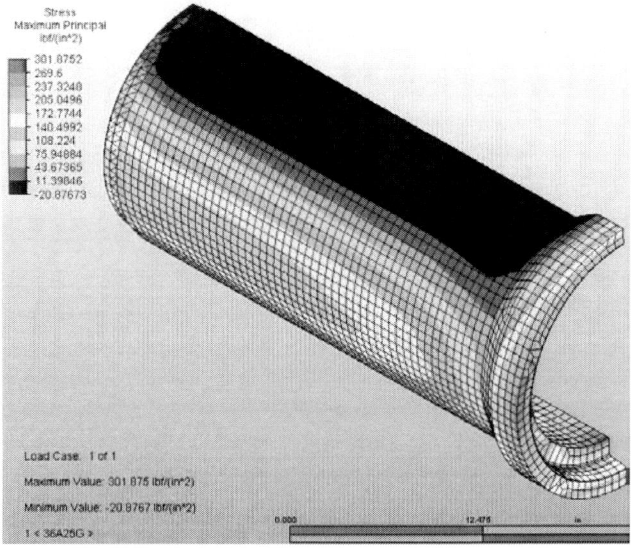

FIG. 6 Results for 36A25 without gasket (82 psi).

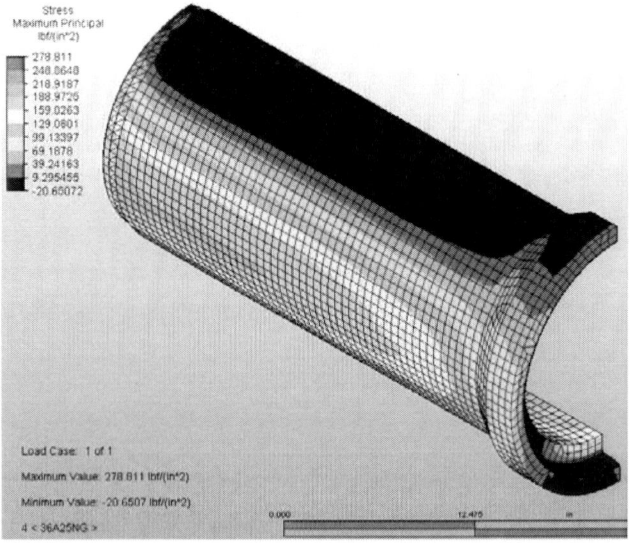

FIG. 7 Results for 36A100 with gasket (263 psi).

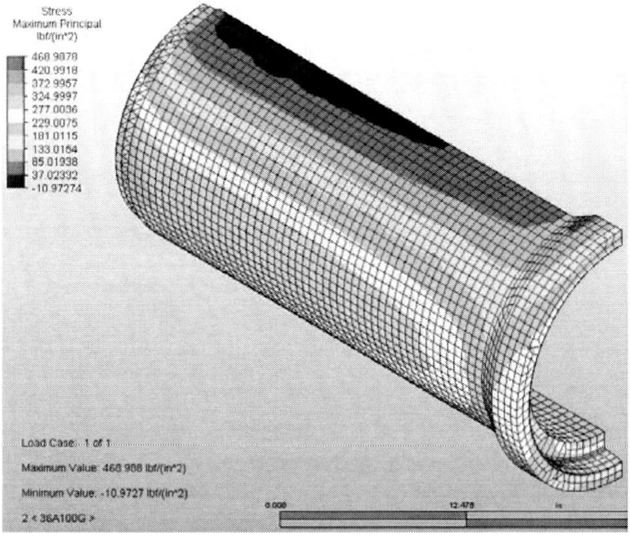

FIG. 8 Results for 36A100 without gasket (148 psi).

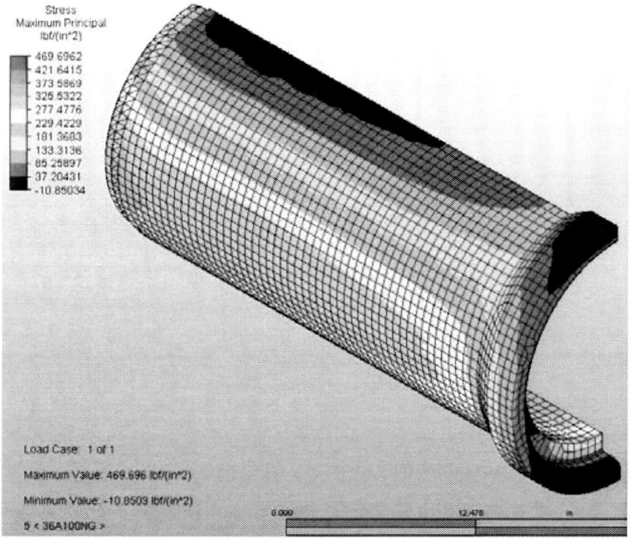

FIG. 9 Results for 36D25 with gasket (484 psi).

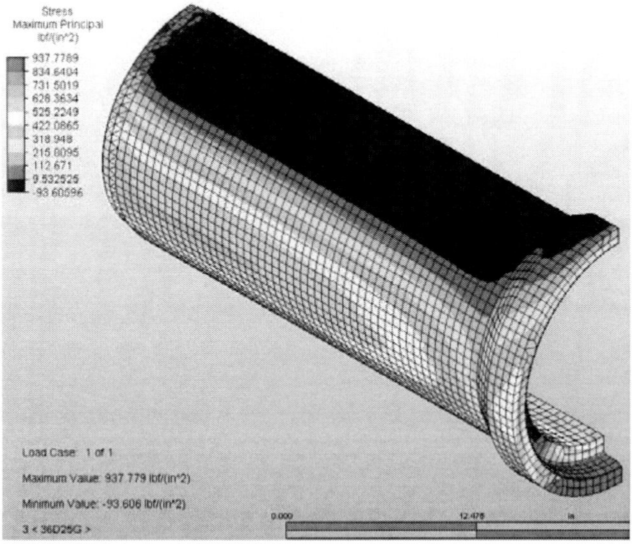

FIG. 10 Results for 36D25 without gasket (337 psi).

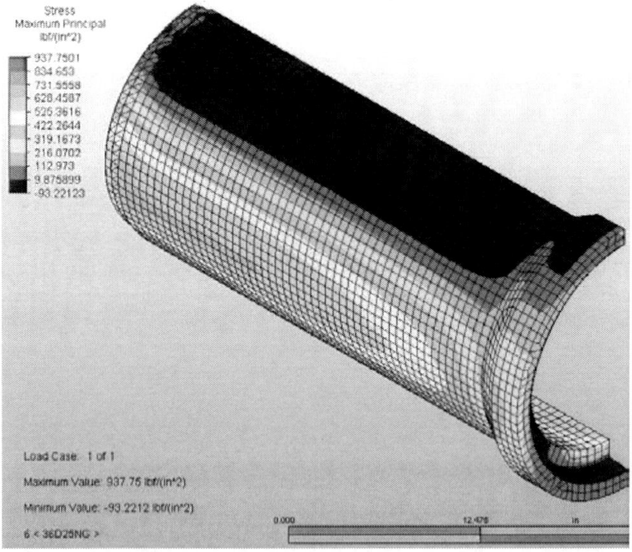

FIG. 11 Results for 36D100 with gasket (496 psi).

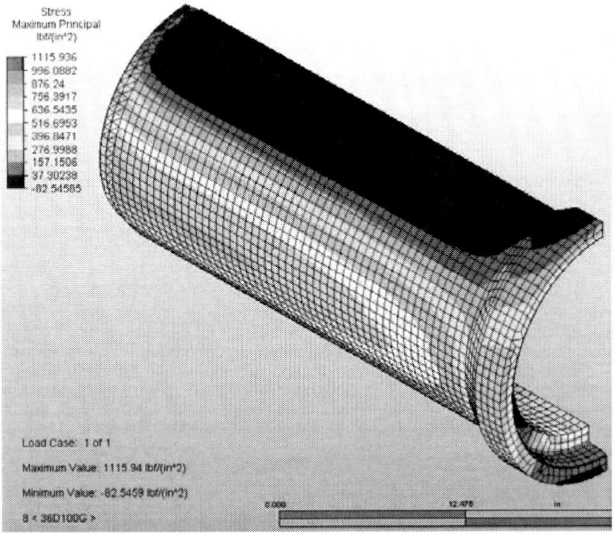

FIG. 12 Results for 36D100 without gasket (388 psi).

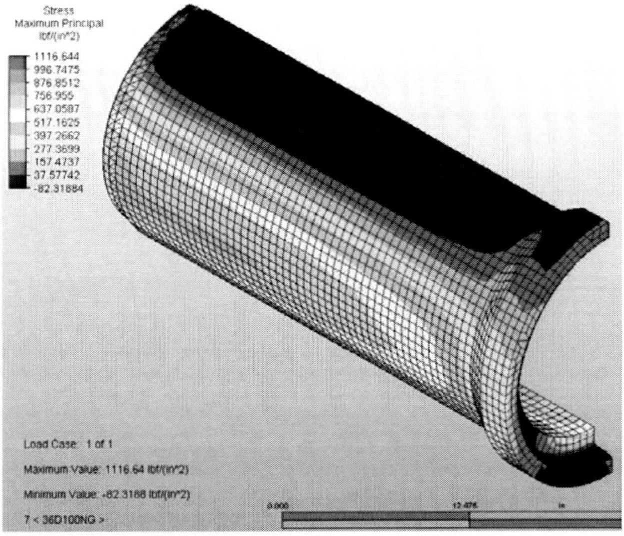

FIG. 13 Results for 72A25 with gasket (402 psi).

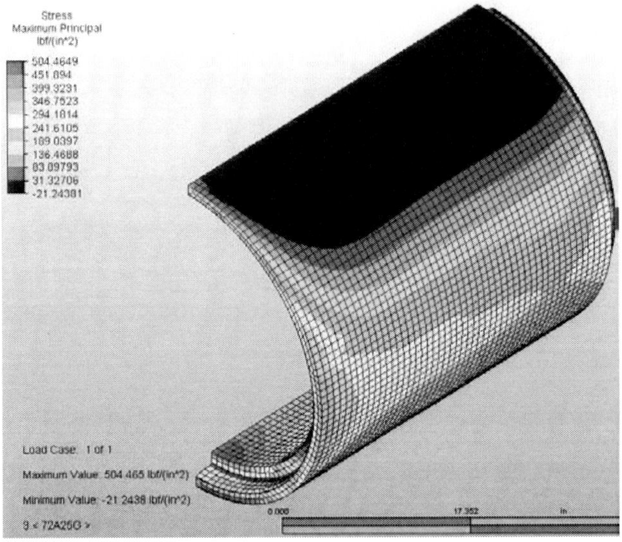

FIG. 14 Results for 72A25 without gasket (236 psi).

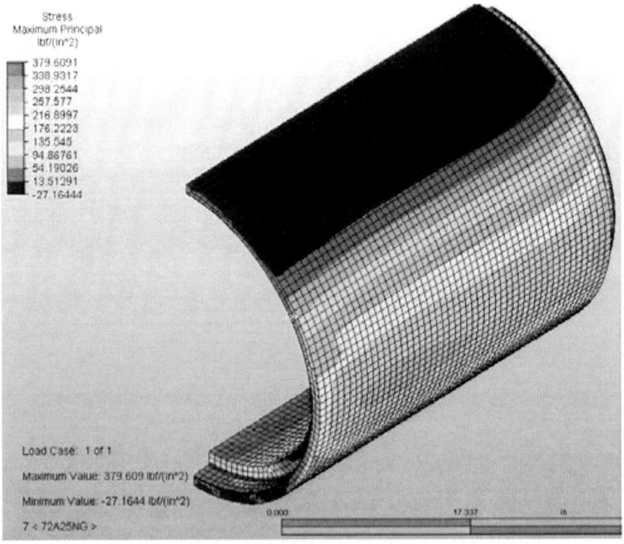

FIG. 15 Results for 72A100 with gasket (483 psi).

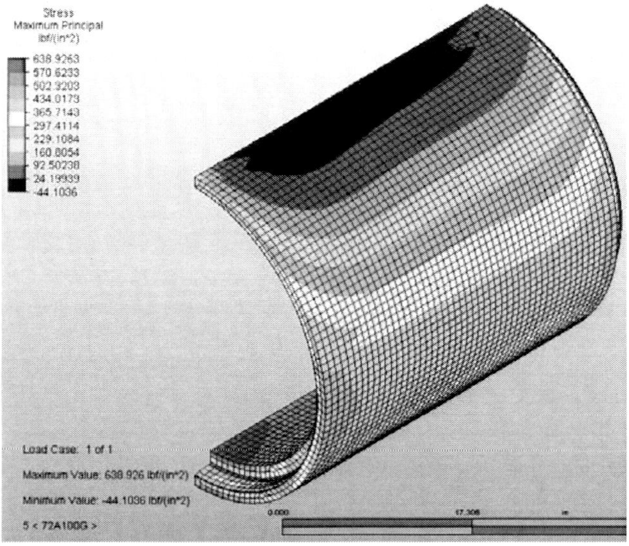

FIG. 16 Results for 72A100 without gasket (354 psi).

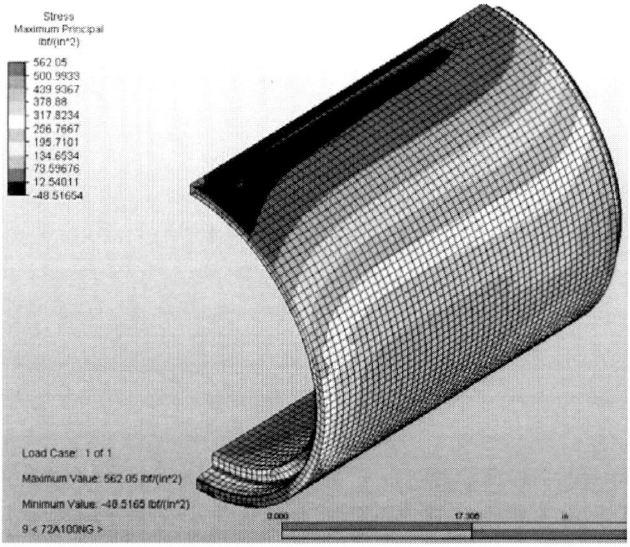

FIG. 17 Results for 72D25 with gasket (935 psi).

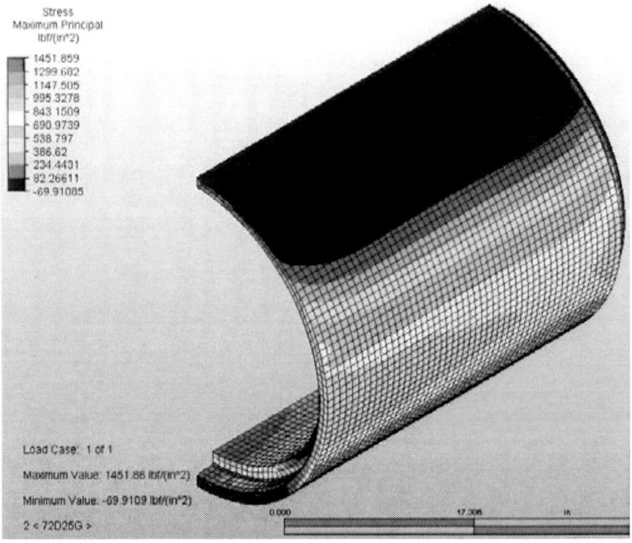

FIG. 18 Results for 72D25 without gasket (826 psi).

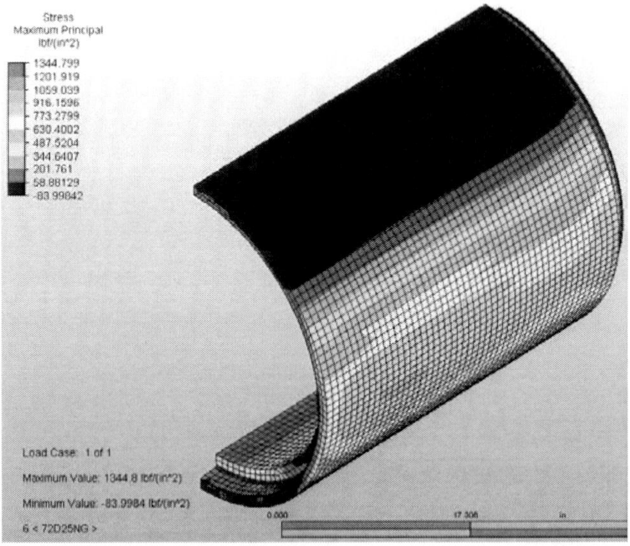

FIG. 19 Results for 72D100 with gasket (1,044 psi).

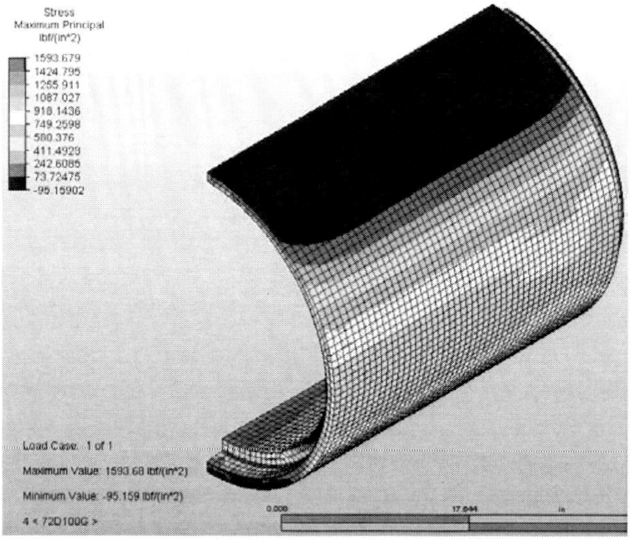

FIG. 20 Results for 72D100 without gasket (945 psi).

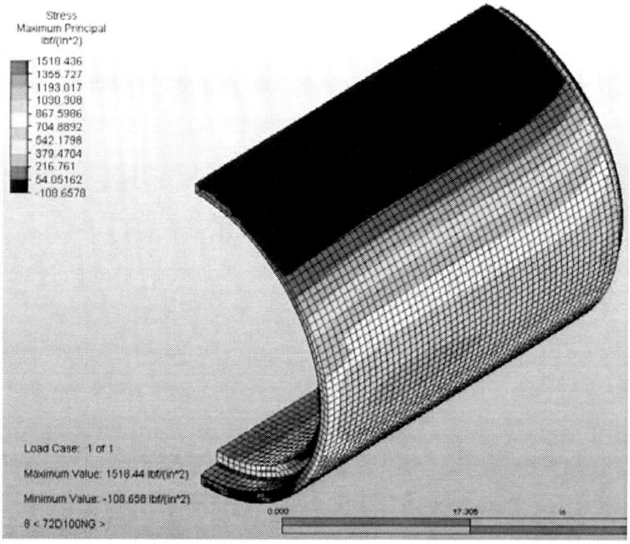

can actually be more than the wall section when the gasket is included. In these sections, the bell does not provide additional stiffness or an increased section. The thinner bell section provides a less stiff section that then sheds a portion of the load to the pipe wall when a gasket is not included. When the gasket is added to the analysis, the stresses in the bell are higher than the stresses in the pipe wall. As the stress plots show, the spigot end of the pipe acts as a stiffened member or edge beam thus increasing the stresses at spring line while the opposite is true of the bell end.

By comparing the aforementioned figures for a given class with and without a gasket, **Table 3** shows the impact of the gasket on the stress in the bell.

The increase in stress caused by the inclusion of a gasket ranges from approximately 100 psi to 170 psi. The increase in stress caused by a change in internal pressure from 25 to 100 ft ranges from 12 psi on the 36-in. size to as much as 110 psi on the 72-in. size. These two comparisons illustrate and confirm that both the gasket pressure and the internal pressure on the pipe impact the stresses on the bell and should be included in the design. Comparison of the pipe sizes based on differences in fill height are shown in **Table 4**.

The impact of fill height on the stresses in the bell as shown in **Table 4** is two to four times greater than that of the gasket and internal pressure differences shown in **Table 3**. Thus, the design of the bell reinforcement on RCLHPP should account for the combined impact of gasket pressure, internal pressure, and fill height.

Recommended Design Procedure

In order to account for the combination of gasket pressure, soil load, and internal pressure, we need to answer the following questions:

1. How do we analyze the impact of an extended bell or straight bell?
2. How do we account for the gasket pressure on the bell?
3. What is the stress limit for the bell reinforcement?

To address Question 1, the stress profile at the spring line of the pipe was investigated. Results of the same size, pressure, and fill classes with and without gaskets are shown in the following figures. The stresses shown for the 36A25 with gasket

TABLE 3 FEA results—gasket versus no gasket.

Size and Class	Stress with Gasket	Stress without Gasket	Stress Difference
36A25	251	82	169
36A100	263	148	115
36D25	484	337	147
36D100	496	388	108
72A25	402	236	166
72A100	483	354	129
72D25	935	826	109
72D100	1,044	945	99

TABLE 4 FEA results—bell stress versus fill height, with and without gasket.

Stresses with Gasket				Stresses without Gasket			
Size and Pressure	Stress at 5 ft (A)	Stress at 20 ft (D)	Stress Difference	Size and Pressure	Stress at 5 ft (A)	Stress at 20 ft (D)	Stress Difference
36 25	251	484	233	36 25	82	337	255
36 100	263	496	233	36 100	148	388	240
72 25	402	935	533	72 25	236	826	590
72 100	483	1,044	561	72 100	354	945	591

illustrate that the gasket pressure is significant at this low-fill/low-pressure condition (**Fig. 21**). When the gasket is removed, the stresses shown in **Fig. 22** indicate that the presence of the extended bell serves to reduce the stress on the section.

Similar results are shown (**Figs. 23** and **24**) for the 36D100 class where the presence of the gasket increases the bell stress but is still less than the calculated wall stress based on the stiffer section.

As shown in **Figs. 21** through **24**, the bell stress also varies from a minimum at the inside face to a maximum at the exterior face. In order to estimate these stresses during the design process, the section can be analyzed per ASTM C361 for the given design moment of the wall section with an increased effective depth and the stress level calculated for the inner and outer faces of the bell. These stresses can then be averaged to provide a design stress due to external loads and the effects of gasket pressure added for analysis of the bell reinforcement.

FIG. 21 Analysis of 36A25 with gasket at spring line.

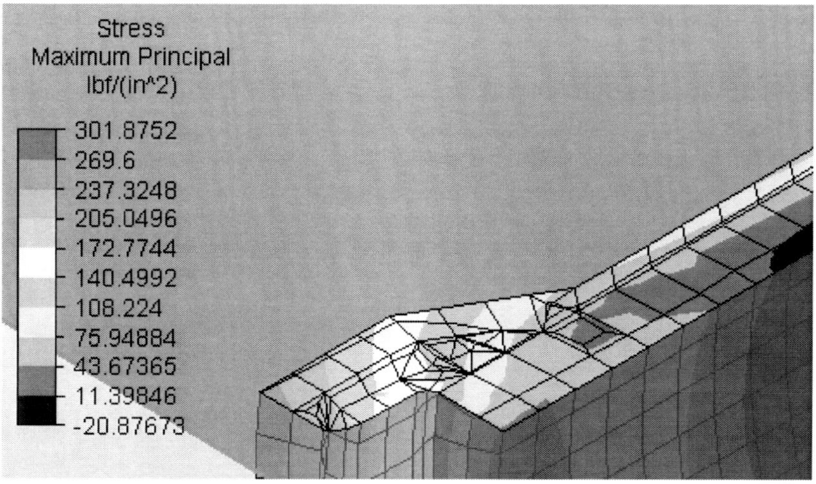

FIG. 22 Analysis of 36A25 without gasket at spring line.

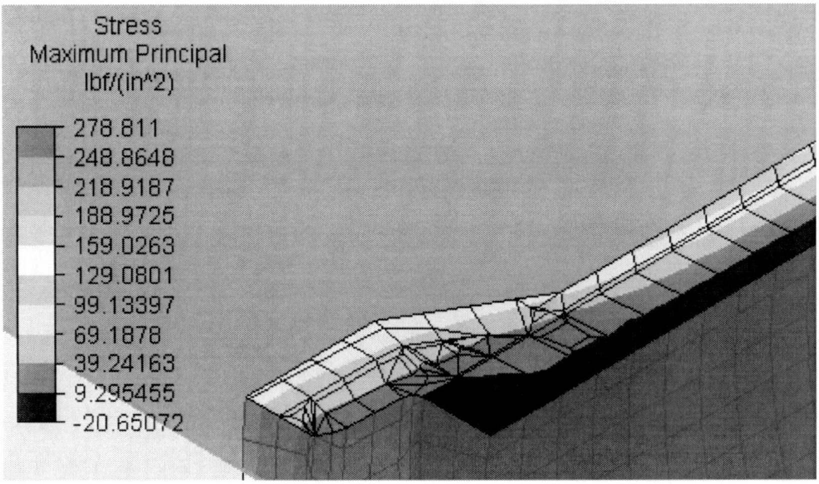

For straight bell sections, **Figs. 25** through **28** illustrate that the thinner bell section will be subjected to similar or higher stresses than the full-wall section. Without the gasket, the maximum stress in the bell is very similar to the full-wall section; when the gasket is added, this stress increases.

FIG. 23 Analysis of 36D100 with gasket at spring line.

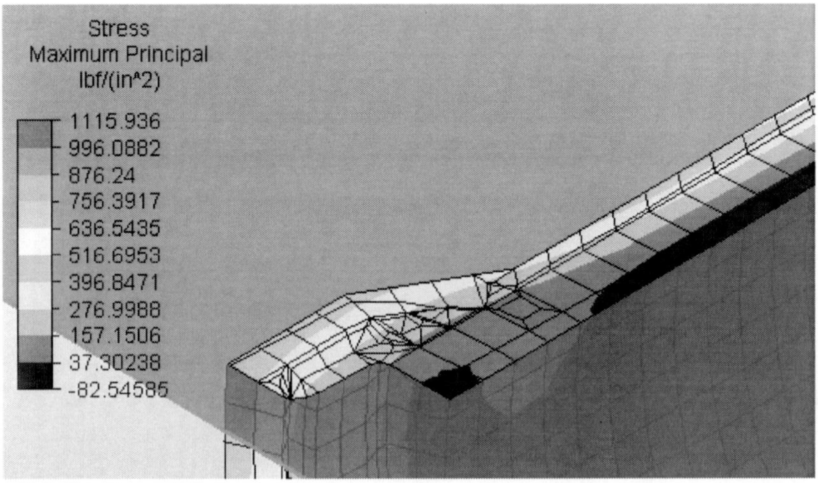

FIG. 24 Analysis of 36D100 without gasket at spring line.

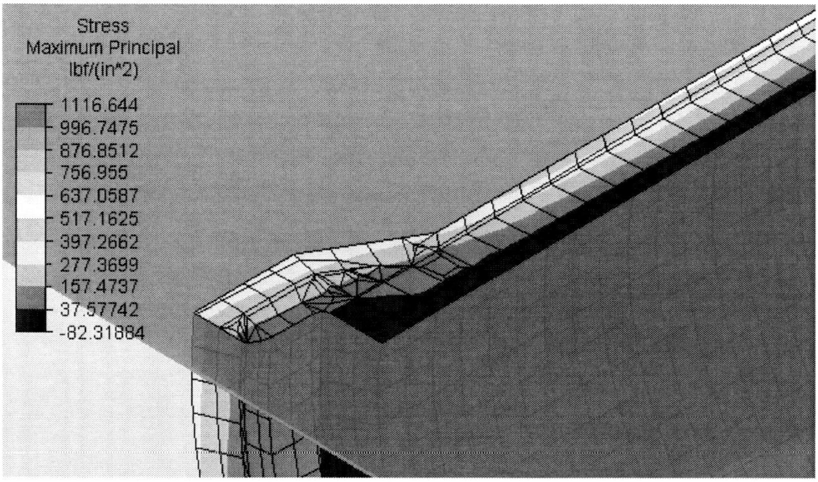

A similar approach as was presented for the 36-in. case is recommended for analysis of a thin bell section. The stresses at the inner (minimum) and outer (maximum) faces of the bell vary and can be determined from the design moment and resulting stresses at those locations. These stresses can then be averaged to provide

FIG. 25 Illustration of 72A25 with gasket at spring line.

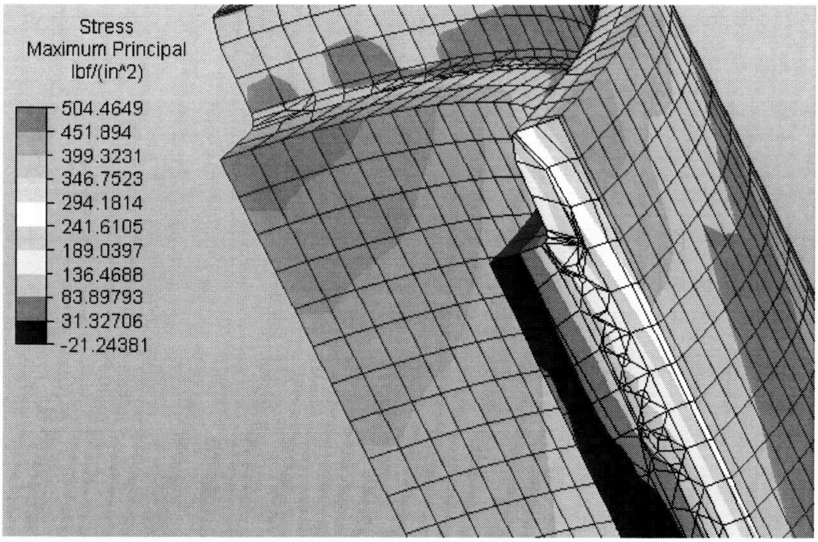

FIG. 26 Illustration of 72A25 without gasket at spring line.

the stress on the bell from external loads. The gasket pressure would then be added to this for analysis of the bell reinforcement.

For consideration, the following procedure is suggested for analysis of RCLHPP bell sections to account for the effects of soil loads, internal pressure, and gasket forces:

FIG. 27 Illustration of 72D100 with gasket at spring line.

FIG. 28 Illustration of 72D100 without gasket at spring line.

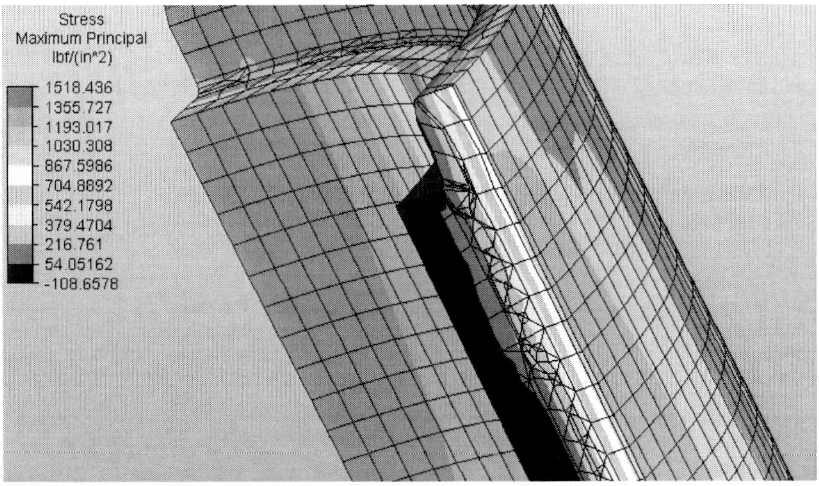

EXAMPLE CALCULATION FOR 36A25

Step 1: Determine the bell stress caused by the gasket pressure (OD = outer diameter).

1. Determine the bell internal/external and pipe inside diameters.

$$\text{ID} = 43\,\text{in.}/\text{OD} = 50.25\,\text{in.}/\text{DIA} = 36\,\text{in. (given by manufacturer)}$$

2. Calculate the bell thickness.

$$\text{BELL} = (0.5) * (\text{OD}-\text{ID}) = 3.625\,\text{in.}$$

3. Determine the wall thickness.

$$\text{WALL} = 4\,\text{in. (given by manufacturer)}$$

4. Calculate the gasket force (Fg) at 200 lbf/in.

$$\text{Fg} = 200\,\text{lbf/in.} * \text{ID} * 0.5 = 4,298\,\text{lb}$$

5. Determine the length (L) of the bell joint.

$$\text{L} = 4.125\,\text{in. (given by manufacturer)}$$

6. Calculate the bell area based on 1.75 multiplied by joint length.

$$\text{Area} = \text{L} * \text{BELL} + 0.75 * \text{L} * \text{WALL} = 27.33\,\text{sq . in.}$$

7. Calculate the bell stress (Θg) from gasket pressure.

$$\Theta\text{g} = \text{Fg}/\text{area} = 157\,\text{psi}$$

Step 2: Determine the bell stress caused by soil and internal pressure per ASTM C361.

1. Calculate the factored moment from gravity loads (Mu).

$$\mathrm{Mu} = 6,787\ \mathrm{in.} * \mathrm{lb/ft}$$

2. Calculate the factored thrust from gravity loads (Nu).

$$\mathrm{Nu} = 1,639\ \mathrm{lb/ft(compression)}$$

3. Calculate the factored thrust from internal pressure (Np).

$$\mathrm{Np} = 2,338\ \mathrm{lb/ft(tension)}$$

4. Calculate the bell thickness (d1).

$$\mathrm{d1} = \mathrm{BELL} = 3.625\ \mathrm{in.}$$

5. Calculate the effective section depth (d2).

$$\mathrm{d2} = 0.5 * (\mathrm{OD} - \mathrm{DIA}) = 7.125\ \mathrm{in.}$$

6. Calculate the effective width factor (ω).

$$\omega = 1.75 * \mathrm{L}/12\ \mathrm{in.} = 0.60$$

7. Calculate the effective section properties.

$$\mathrm{I} = 0.0833 * (1.75 * \mathrm{L}) * (\mathrm{d2})\hat{}3$$
$$\mathrm{I} = 217.6\ \mathrm{in.}\hat{}4$$
$$\mathrm{yo} = 0.5 * \mathrm{d2} = 3.56\ \mathrm{in.}$$
$$\mathrm{yi} = \mathrm{yo} - \mathrm{d1} = -0.0625\ \mathrm{in.}$$

8. Calculate the bell stress at the OD (Θo).

$$\Theta\mathrm{o} = \mathrm{Mu} * \omega * \mathrm{yo/I} = 66.8\ \mathrm{psi}$$

9. Calculate the bell stress at the ID (Θi).

$$\Theta\mathrm{i} = \mathrm{Mu} * \omega * \mathrm{yi/I} = -1.2\ \mathrm{psi}$$

10. Calculate the average bell stress from gravity loads (Θave).

$$\Theta\mathrm{ave} = 0.5 * (\Theta\mathrm{o} + \Theta\mathrm{i}) - \mathrm{Nu} * \omega/\mathrm{area} = -3.2\ \mathrm{psi}$$

11. Calculate the bell stress from internal pressure (Θint).

$$\Theta\mathrm{int} = \mathrm{Np} * \omega/\mathrm{Area} = 51.5\ \mathrm{psi}$$

12. Calculate the design bell stress (Θdes).

$$\Theta\mathrm{des} = \Theta\mathrm{Ave} + \Theta\mathrm{int} = 48.2\ \mathrm{psi}$$

Step 3: Determine the bell reinforcement required.

1. Determine the final combined bell stress (Θfinal).

$$\Theta final = \Theta des + \Theta g = 206 \text{ psi}$$

2. Determine the allowable steel stress (Θallow).

$$\Theta allow = 17,000 - 35 * HEAD = 16{,}125 \text{ psi}$$

3. Calculate the bell reinforcement (As).

$$As = Area * \Theta final / \Theta allow = 0.35 \text{ sq.in.}$$

The previous example illustrates the average stress for an extended bell. Based on the FEA model results, the response of the stiffened section at the bell is expected to reduce the steel area required. As the pipe size gets larger, the extended bell is no longer provided by most suppliers. As was shown in the FEA model, a smooth bell illustrates an increase in steel area required. Following is the same analysis for a straight bell with a 72-in. diameter:

EXAMPLE CALCULATION FOR 72A25

Step 1: Determine the bell stress caused by the gasket pressure.

1. Determine the bell internal/external and pipe inside diameter.

$$ID = 79.25 \text{ in.}/OD = 86 \text{ in.}/DIA = 72 \text{ in. (given by manufacturer)}$$

2. Calculate the bell thickness.

$$BELL = (0.5) * (OD - ID) = 3.375 \text{ in.}$$

3. Determine the wall thickness.

$$WALL = 7 \text{ in. (given by manufacturer)}$$

4. Calculate the gasket force at 200 lbf/in.

$$Fg = 200 \text{ lbf/in.} * ID * 0.5 = 7,925 \text{ lb}$$

5. Determine the length of the bell joint.

$$L = 5.25 \text{ in. (given by manufacturer)}$$

6. Calculate the bell area based on 1.75 multiplied by joint length.

$$Area = L * BELL + 0.75 * L * WALL = 45.28 \text{ sq.in.}$$

7. Calculate the bell stress from gasket pressure.

$$\Theta g = Fg/Area = 175 \text{ psi}$$

Step 2: Determine the bell stress caused by soil and internal pressure per ASTM C361.

1. Calculate the service moment from gravity loads.

$$\text{Mu} = 30,367 \text{ in.} * \text{lb/ft}$$

2. Calculate the service thrust from gravity loads.

$$\text{Nu} = 3,075 \text{ lb/ft (compression)}$$

3. Calculate the service thrust from internal pressure.

$$\text{Np} = 4,676 \text{ lb/ft (tension)}$$

4. Calculate the bell thickness.

$$\text{d1} = \text{BELL} = 3.375 \text{ in.}$$

5. Calculate the effective section depth.

$$\text{d2} = 0.5 * (\text{OD} - \text{DIA}) = 7 \text{ in.}$$

6. Calculate the effective width factor.

$$\omega = 1.75 * \text{L}/12 \text{ in.} = 0.766$$

7. Calculate the effective section properties.

$$\text{I} = 0.0833 * (1.75 * \text{L}) * (\text{d2})\hat{}3$$
$$\text{I} = 263 \text{ in.}\hat{}4$$
$$\text{yo} = 0.5 * \text{d2} = 3.5 \text{ in.}$$
$$\text{yi} = \text{yo} - \text{d1} = 0.125 \text{ in.}$$

8. Calculate the bell stress at the OD.

$$\Theta\text{o} = \text{Mu} * \omega * \text{yo/I} = 310 \text{ psi}$$

9. Calculate the bell stress at the ID.

$$\Theta\text{i} = \text{Mu} * \omega * \text{yi/I} = 11 \text{ psi}$$

10. Calculate the average bell stress from gravity loads.

$$\Theta\text{ave} = 0.5 * (\Theta\text{o} + \Theta\text{i}) - \text{Nu} * \omega/\text{Area} = 108 \text{ psi}$$

11. Calculate the bell stress from internal pressure.

$$\Theta\text{int} = \text{Np} * \omega/\text{Area} = 79 \text{ psi}$$

12. Calculate the design bell stress.

$$\Theta\text{des} = \Theta\text{Ave} + \Theta\text{int} = 188 \text{ psi}$$

Step 3: Determine the bell reinforcement required.

1. Determine the final combined bell stress.

$$\Theta\text{final} = \Theta\text{des} + \Theta\text{g} = 363\,\text{psi}$$

2. Determine the allowable steel stress.

$$\Theta\text{allow} = 17,000 - 35 * \text{HEAD} = 16{,}125\,\text{psi}$$

3. Calculate the bell reinforcement.

$$\text{As} = \text{area} * \Theta\text{final}/\Theta\text{allow} = 1.02\,\text{sq.in.}$$

Compare Recommended Design Procedure with USBR M-1 Reinforcing Areas

Analysis of the bell reinforcement required by USBR M-1 as compared to the areas calculated per the proposed method outlined earlier is presented in **Table 5**. Bell reinforcement according to the USBR M-1 standard is shown the same in both tables for illustration purposes.

The calculations would indicate that a significant increase is needed for all sizes, with the increase being bigger for larger diameter sections. This may be due to an inherent oversight in the current design procedure. The bell reinforcement shown in the USBR M-1 standard by specification is based only on internal pressure acting on a 1.75-in. length of the bell and the gasket force applied to the bell. The resulting steel area is based strictly on the tensile force generated in the bell by these two forces. But, as the FEA results have shown, the bell is subjected to significant stresses caused by the internal and external loading of the pipe wall. One potential way to address this would be to include both the steel area shown in **Table 2** of the USBR M-1 standard for a given fill height and pressure class and to *add* the area shown in **Table 1** of the USBR M-1 standard to get a combined reinforcement required. **Table 6** illustrates the resulting steel areas and how they compare to the proposed design method.

Comparing the combined USBR M-1 **Tables 1** and **2** steel areas to the areas calculated by the proposed method shows a much better agreement between the two

TABLE 5 Comparison between USBR M-1 and proposed design method.

Bell Reinforcement at 200 lbf/in.				Bell Reinforcement at 150 lbf·in.			
Size and Class	USBR M-1 Std	Proposed Method	% Increase	Size and Pressure	USBR M-1 Std	Proposed Method	% Increase
36A25	0.292	0.348	19 %	36A25	0.292	0.282	−3 %
36D100	0.439	0.665	51 %	36D100	0.439	0.586	33 %
72A25	0.538	1.02	90 %	72A25	0.538	0.895	66 %
72D100	0.81	2.621	224 %	72D100	0.81	2.474	205 %

TABLE 6 Combined USBR M-1 Tables 1 and 2 areas compared with proposed method.

Combined USBR M-1 Tables 1 and 2—200 lbf/in.			
Size and Class	USBR M-1 Std	Proposed Method	% Increase
36A25	0.42	0.348	−17 %
36D100	0.91	0.665	−27 %
72A25	0.84	1.02	21 %
72D100	1.86	2.621	41 %

sets of assumptions, as shown in **Table 6**. The proposed method should serve as a very conservative upper bound solution to the structural design of the reinforcement in the bell. For the lower bound of the solution, the pipelines that have been fabricated and installed according to the existing USBR M-1 standard, incorporating the bell reinforcement as shown in **Table 2** of that standard, have a satisfactory performance history. The assumptions included in the USBR M-1 standard (200 lbf/in.) likely have an adequate factor of safety built into the system because the gasket pressure is not constant on the entire perimeter of the joint. Production tolerances vary such that the pressure may only be maximum over a small portion of the perimeter. Rationalizing this variability is not feasible for design.

The joint integrity in a line of RCLHPP is critical to the longevity of the line and to the performance of the system. The joint is only a very small portion of the overall mass of the structure (typically 10 % or less of the overall length of each piece), but it has a large impact on the overall success of an installation. The incremental cost of providing a more robust joint is relatively inconsequential compared to rehabilitation costs. With the advent of the updated ASTM C361 design process in 2011, a more rational design method for the joint reinforcement naturally follows. One option is presented within for consideration. This option, however, has not been verified with field testing and calibration of the model. Stresses caused by joint shear from nonuniform longitudinal support and other external factors not addressed by this paper should also be incorporated to encapsulate the variables necessary for a complete design process.

Conclusions

The design standard provided by ASTM C361 currently provides guidance on reinforcement of the wall of a concrete pressure pipe and has replaced the USBR M-1 standard. ASTM C361 does not provide guidance on structural design of the joints as was included in the USBR M-1 standard. A series of FEA models were analyzed to determine stress levels in the concrete bell of a typical USBR R-3 or R-4 joint as detailed by the USBR M-1 standard. The results of the FEA models illustrate that the bell acts in tandem with the pipe wall and is exposed to similar stresses under normal loading conditions. The joint configuration provided as extended bell or smooth bell had a large impact on the stress response to load. These models were analyzed to develop a recommended design procedure to incorporate the soil loading, internal pressure, and gasket pressures on the joint. This design procedure presented within is a potential option but should be verified with full-size specimen calibration for the model.

References

[1] USBR M-1, *Standard Specifications for Reinforced Concrete Pressure Pipe*, U.S. Bureau of Reclamation, Washington, DC, 1991.

[2] ASTM C361, *Standard Specification for Reinforced Concrete Low-Head Pressure Pipe*, ASTM International, West Conshohocken, PA, 2014, www.astm.org

[3] AWWA C302, *Reinforced Concrete Pressure Pipe, Noncylinder Type*, American Water Works Association, Denver, CO, 2011, www.awwa.org

[4] AWWA M9, *Concrete Pressure Pipe*, 3rd ed., American Water Works Association, Denver, CO, 2008, www.awwa.org

STP 1601, 2017 / available online at www.astm.org / doi: 10.1520/STP160120160122

Pardeep Sharma[1]

Evolution of Precast Box Culvert Joint and Sealing

Citation

Sharma, P., "Evolution of Precast Box Culvert Joint and Sealing," *Concrete Pipe and Box Culverts, ASTM STP1601*, J. J. Meyer and J. Beakley, Eds., ASTM International, West Conshohocken, PA, 2017, pp. 118–134, http://dx.doi.org/10.1520/STP160120160122[2]

ABSTRACT

This paper will illustrate the evolution of box culvert joints and sealing. Not every box culvert project contains the identical level of watertight requirements; some box culverts are exposed on each end of the water transportation through the section. For this type of application, butyl sealant has been used. These units must be sealed, but hydrostatic watertightness is not crucial. Certain box culverts need to be regulated for erosion over the joint. For erosion control applications, an exterior joint wrap and butyl sealant have been utilized. On the other hand, several box culverts are used as water containment structures. Presently, the issue has been to assess the precast box culvert system. Over the years, the precast box culvert specification has been revised or changed completely; due to the watertight system requirements, ASTM C1677, *Standard Specification for Joints for Concrete Box, Using Rubber Gaskets*, has established. This specification incorporates flexible joints for the concrete box section and the application of rubber gaskets for leak-resistant joints.

Keywords

evolution, box culvert joint, gasketed

Description and Historical Overview

Concrete box culverts are ever-present; they have been used since the beginning of the twentieth century and are still in use today. The first standard plans for roadway

---

Manuscript received September 5, 2016; accepted for publication November 18, 2016.
[1]Hamilton Kent LLC, 740 Baxter Ln., Winchester, TN 37398
[2]ASTM Symposium on *Concrete Pipe and Box Culverts* on December 7, 2016 in Orlando, Florida.

structures, which included designs for both "box culverts" and "box bridges" were issued in 1912. The culvert ranged from 18 in. by 18 in. to 6 ft by 6 ft and included reinforced concrete on all four sides of the box [1].

Between 1935 and 1945, the Maryland State Roads Commission reports contained numerous references to the construction of box culverts utilized on state roadways [1]. Generally, these boxes were monolithic pour, and sealing was not a concern with cast-in-place boxes. Precast boxes, which have been in use for 50 years, initially followed ASTM C789, *Standard Specification for Precast Reinforced Concrete Box Sections for Culverts, Storm Drains, and Sewers.* Typically, a box culvert is a four-sided single-cell structure with square or rectangular openings (see **Fig. 1** and **Fig. 2**).

Fig. 3 is an example of multiple cell squares or rectangular culverts. **Figs. 4–6** are examples of segmental precast reinforced concrete box sections.

ASTM Standards for Precast Box Culverts

- ASTM C850-00, *Standard Specification for Precast Reinforced Concrete Box Sections for Culverts, Storm Drains, and Sewers with Less Than 2 ft of Cover*

FIG. 1 Single-cell square culvert, not typical.

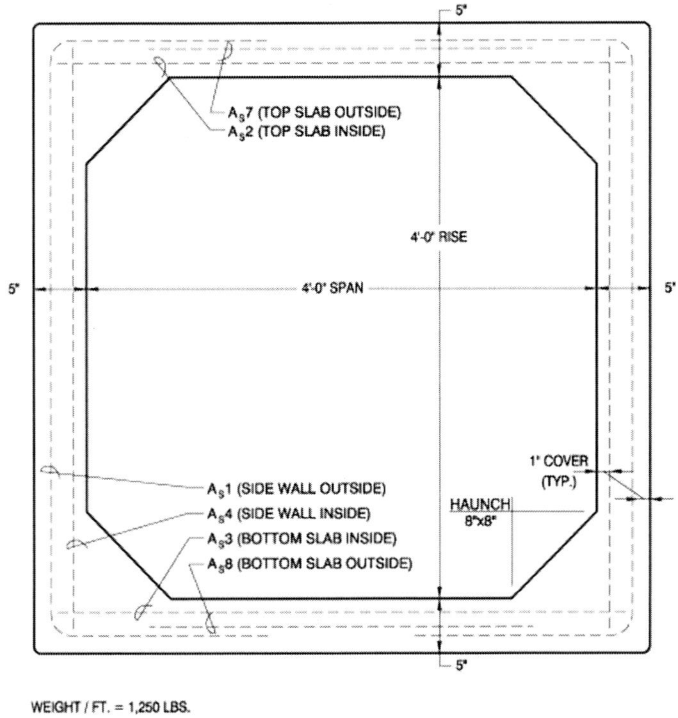

FIG. 2 Single-cell rectangular culvert, not typical.

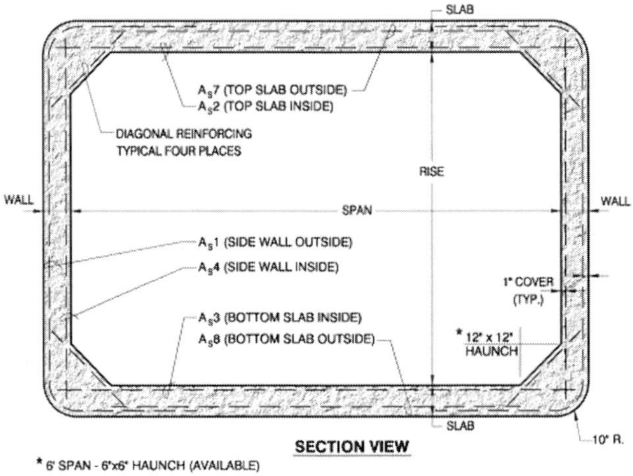

FIG. 3 Multiple cell culverts.

FIG. 4 Three-sided rigid frame with footer.

Subjected to Highway Loadings, was withdrawn in 2000 and replaced by ASTM C1433, *Standard Specification for Precast Reinforced Concrete Monolithic Box Sections for Culverts, Storm Drains, and Sewers* [2].

- ASTM C789-00: This specification covers single-cell, precast concrete box sections intended to be used for the construction of culverts for the conveyance of storm water, industrial wastes, and sewage. It was withdrawn in 2000 and replaced by ASTM C1433 [2].
- ASTM C1433-16a: This specification covers single–cell, precast, reinforced concrete box sections cast monolithically and intended to be used for the

FIG. 5 Three-sided, U-shaped culvert with a flat-top slab.

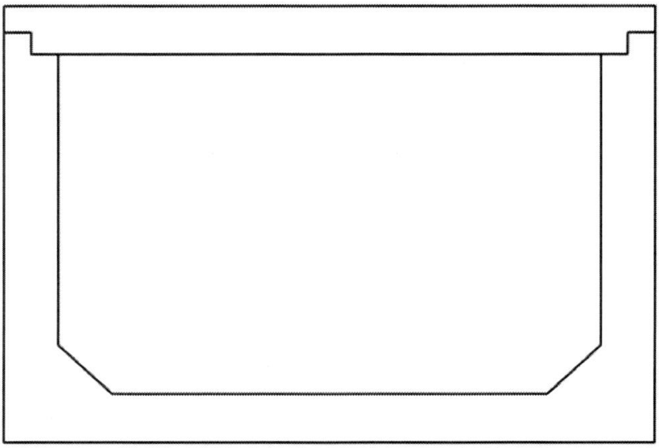

FIG. 6 Segmental, precast, reinforced concrete box sections.

construction of culverts and for the conveyance of storm water industrial wastes and sewage. This standard was adopted in 2000, replacing both ASTM C850 and ASTM C789 [2].

- ASTM C1504-16, *Standard Specifications for Manufacture of Precast Reinforced Concrete Three-Sided Structures for Culverts and Storm Drains,* covers single-cell, precast, conventionally reinforced concrete three-sided structures intended to be used for the constructions of culverts and for the conveyance of storm water [2].
- ASTM C1577-16, *Standard Specification for Precast Reinforced Concrete Monolithic Box Sections for Culverts, Storm Drains, and Sewers Designed According to AASHTO LRFD,* covers single-cell, precast, reinforced concrete box sections cast monolithically and intended to be used for the construction of culverts and for the conveyance of storm water, industrial waste, and sewage [2].
- ASTM C1677-11a, *Standard Specification for Joints for Concrete Box, Using Rubber Gaskets,* covers flexible joints for concrete box sections, using rubber gaskets for leak-resistant joints as well as the design of joints and the requirements for rubber gaskets to be used therewith, for boxes conforming in all other respects to ASTM C1433 or ASTM C1577 provided that, if there is conflict in permissible variations in dimensions, the requirements of this specification for joints govern [2].
- ASTM C1786-16, *Standard Specification for Segmental Precast Reinforced Concrete Box Sections for Culverts, Storm Drains, and Sewers Designed According*

to AASHTO LRFD, covers precast, reinforced, concrete box sections comprised of separate segments that, once properly field assembled, make the final structure. These structures are intended to be used for the construction of culverts and for the conveyance of storm water, industrial wastes, and sewage. ASTM C1786 is the first ASTM standard specifically covering precast segmental box culverts [2].

Precast Box Culvert Manufacturing

Most manufacturers use the wet casting method for precast box culvert construction. In this method, the form is assembled and then concrete is pumped into the form and left to cure for a minimum of 4 h. Alternatively, some manufacturers use what is known as the dry cast method, which requires the use of mechanical equipment. The cast product is immediately stripped, and the form is then reused. Single-form and modular tooling (**Fig. 7**) is available for the manufacturing of box culverts. **Fig. 8** shows the interior form and stripping of the outer form.

Evolution of Joints

Following are figures demonstrating the evolution of box culvert joints. **Figs. 9–14** shows the transformation from tongue and groove to single offset joints.

Evolution of Box Culvert Sealing

The joints that connect the precast box sections are constantly problematic. Frequently, issues concerning joints have been centered on the joint fit-up or the

FIG. 7 Example of modular tooling and equipment.

FIG. 8 Inner form and stripping.

jointing material. This can be rectified by having the precast box culvert manufacturer verify the fit of the joint.

BUTYL SEALANT

Butyl sealant (**Fig. 15**) is used to seal the box culvert on the tongue and grove of the joint.

FIG. 9 Example of keyway joint, may not be typical.

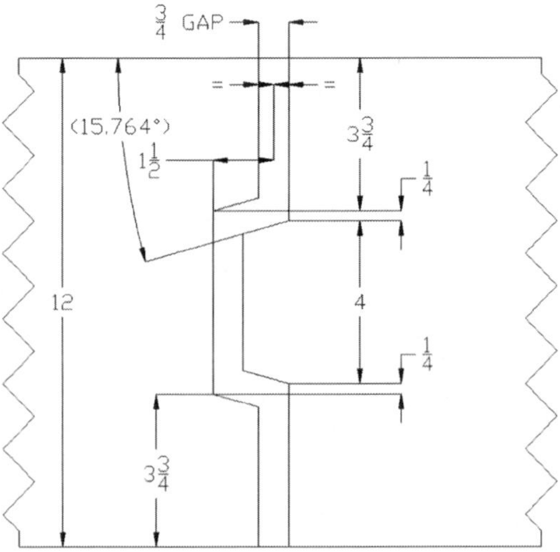

FIG. 10 Tongue and groove/keyway joint (10° to 15° slope and greater).

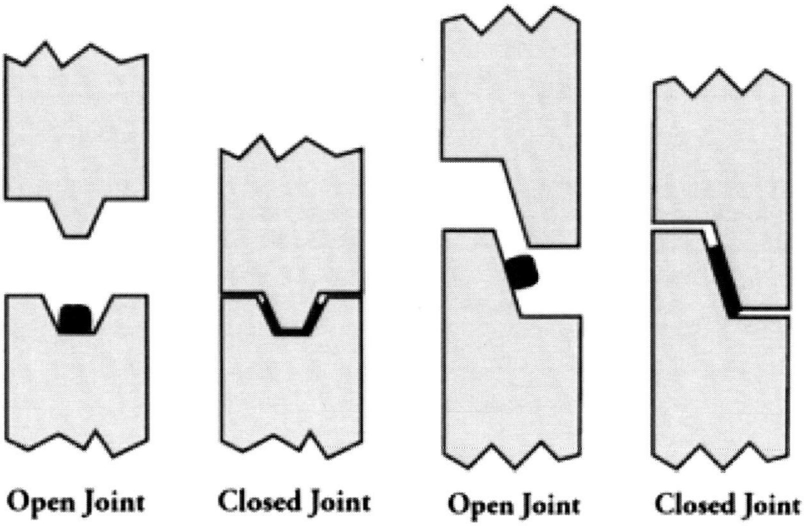

FIG. 11 Typical tongue and groove joint.

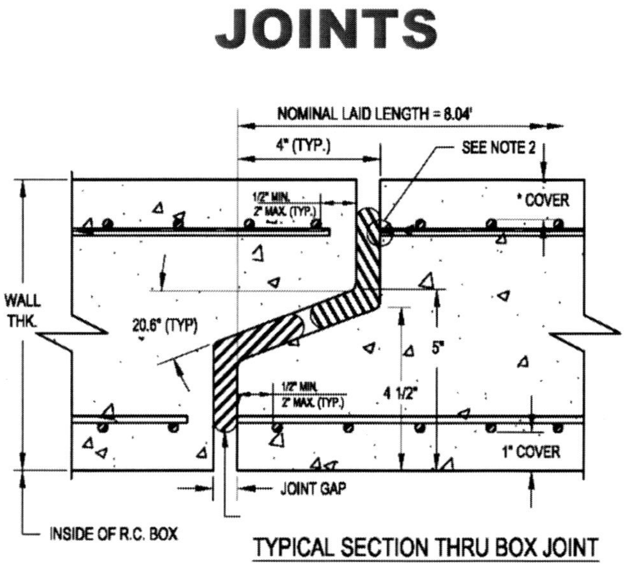

FIG. 12 Example of tongue and groove joint (less than 5° slope), may not be typical.

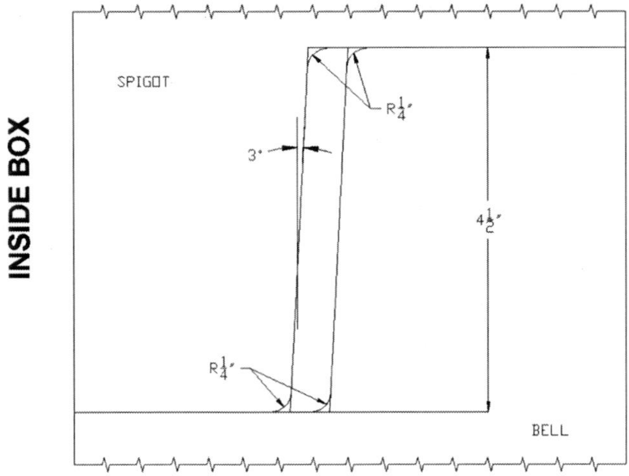

FIG. 13 Example of single offset gasketed joint—single form, may not be typical.

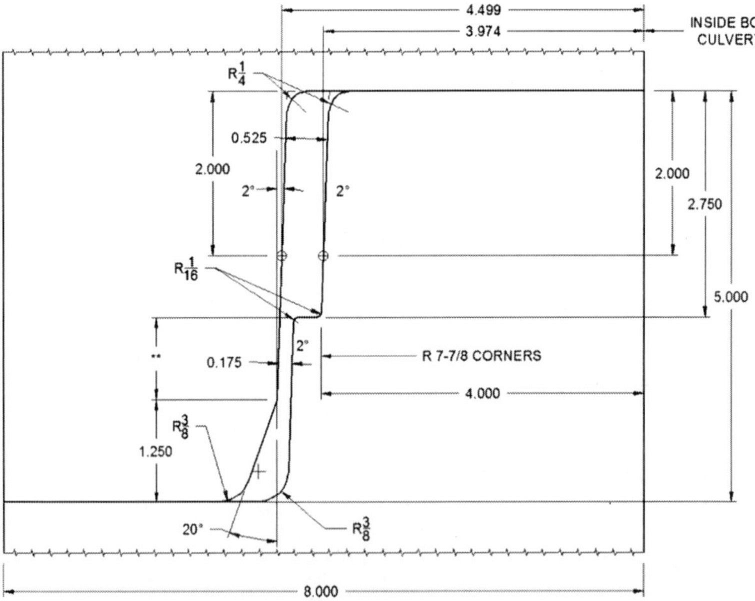

FIG. 14 Example of single offset gasketed joint—modular tooling may not be typical.

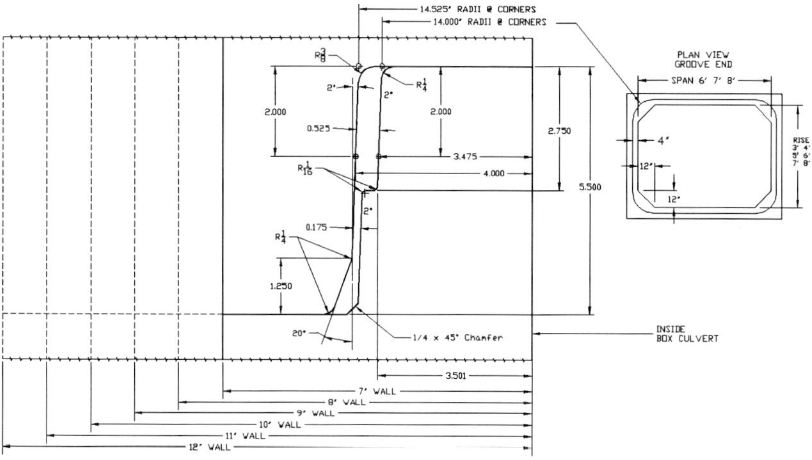

Preformed butyl sealant is used as filler, which then allows for a suitable seal. During assembly of the joint, the sealant is compressed, causing displacement. As the sealant is compressed, the resistant force for further compression increases as a function of the percentage of compression. **Fig. 16** shows compression on the Y-axis and force on the X-axis. This force compression curve is based on a 0.525-in. total gap on the joint with 1-in. equivalent butyl sealant (height 0.9 in.).

From **Fig. 16**, the force to compress 0.3 in. is 50 lb and to compress 0.4 in. is approximately 125 lb. The increase of compression by 33 % will provide an increase in force by 150 %. In this case, the force for assembly is increased 2.5 times.

FIG. 15 Tongue and groove box culverts.

FIG. 16 Compression versus force curve for butyl sealant.

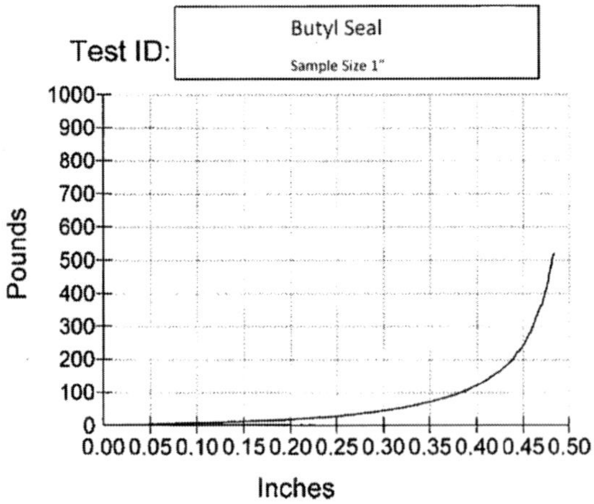

As the sealant is compressed, the sealing gasket increases in width. This width in turn increases the surface area of the applied force. Force compression data show that, when there is an excess of sealant, that excess may cause problems in compressing.

Sealant placement is critical in preventing leakage from the joint. It is required to place sealant in an annular space. Preformed flexible butyl sealants (**Fig. 17**) are required to meet the composition and performance testing requirements of ASTM C990, *Standard Specification for Joints for Concrete Pipe, Manholes, and Precast Box Sections Using Preformed Flexible Joint Sealants.*

Sealant is able to adhere to dry and clean surfaces (**Fig. 18**). To improve adhesion, it is recommended that primer be used to create an ideal surface that allows for sealant placement if the area has been predisposed to moisture or colder temperatures (**Fig. 19**).

EROSION CONTROL, LEAK-RESISTANT SEALING FOR THE TONGUE AND GROOVE JOINT

In addition to sealant, the joints need to be covered with an exterior membrane for prevention of erosion. Sealant is installed as shown in **Fig. 18**; a sealant is not used exclusively—an exterior joint wrap is also applied to the top and sides of the box culvert joint. This exterior wrap is installed after placement, as shown in **Fig. 20**.

For purposes of leak resistance on standard tongue and groove joints, in addition to butyl joint sealant, an expanding water stop sealant can be used on the perimeter of the tongue, and elastomeric material can be caulked or troweled around

FIG. 17 Typical preformed butyl sealant.

FIG. 18 Sealant installed on box culvert.

FIG. 19 Sealant installed using primer.

FIG. 20 Wrap installed on box culvert.

FIG. 21 Gasket installed on a sharp corner.

and onto the interior of the joint. Different types of external bands can be used to provide leak resistance.

GASKET FOR TONGUE AND GROOVE JOINT

The "Achilles heel" of the rubber gasketing system for the box section has always been the inability to produce a gasket seal around the sharp corner. The corner molded gasket, as shown in **Fig. 21**, is compressed between the tongue and groove as the box section homes until the gasket forms a solid seal in the annular space, thereby making it watertight.

Due to the increase in gap as shown in **Fig. 22**, the corner-molded gasket is only recommended for up to a 5° taper on horizontal application.

For a 0.25-in. gap, gasket height at 40 % deformation would be 0.417 in. With an increased gap (0.250 + 0.088 = 0.338 in.), deformation would be only 19 %, resulting in a possible leakage problem.

Corner-molded gaskets for horizontal application are only recommended to meet the specifications of ASTM C1675, *Standard Practice for Installation of Precast Reinforced Concrete Monolithic Box Sections for Culverts, Storm Drains, and Sewers.*

GASKETS FOR SINGLE OFFSET JOINT

The single offset joint design on a box culvert is similar in pipe application. However, due to the rectangular or square shape, it is recommended to use a minimum of 6.00 in. radius on the corner. Because the joint is similar to the pipe, this means that all the benefits associated with a round pipe are the same for the box culvert joint.

For the box culvert function, gaskets are required to be installed at the manufacturing location. **Fig. 23** shows a pre-lube gasket installed on a single offset of the joint. The gasket sags at the lower edge (base) and therefore must be glued.

FIG. 22 Increase in gap with joint opening.

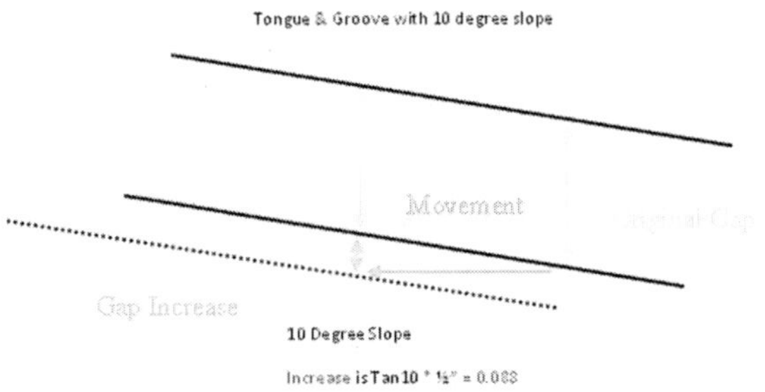

In recent years, several box section manufacturers have begun engineering their structures with a single offset joint, as shown in **Fig. 24**. This joint, along with radius corners, allows for the use of any type of profile gasket. A pre-lubricated gasket is used most often due to the ease of installation and homing of the sections in the field.

The box culvert with installed gaskets can be stored in a yard for a year without any detrimental effects on the gaskets thanks to the ethylene propylene diene monomer (EPDM) material used in the manufacturing of these gaskets (**Fig. 24**).

Gasket design calculations on the box culvert joint are analogous to the concrete round pipe. There are approximately 35 gasketed box culvert producers in North America, and this number has been rising annually. There have been many projects completed in North America that required more stringent joint performance requirements in comparison to ASTM C1677. One of the most recent projects was undertaken by the city of Lubbock, Texas. It required installing a new underground storm water system between two playa lakes. A hydrostatic test was obligatory for the gaskets box section and showed 13 psi for 48 h with no leak. Over the past 20 years,

FIG. 23 23 Pre-lube gasket installed on spigot of box.

FIG. 24 Storage of box culvert with gasket installed on spigot of box.

several other governing agencies, such as the city of New York, have specified performance requirements of box sections to be 13 psi.

It is easier to meet ASTM C1677 or 13 psi requirements (or both) when precision is used. Fig. 25 shows a force deformation curve on a typical single offset joint with a pre-lube gasket. The joint is based on 0.525 in. total gap and 0.175 in. annular space. A test was done based on 15 % stretch on the gasket. The length of the sample is 1.2654 in., and the width of profile is 1.1032 in., which provides a cross-section area of 1.3946 in.

FIG. 25 Force deformation curve on single offset joint with pre-lube gasket.

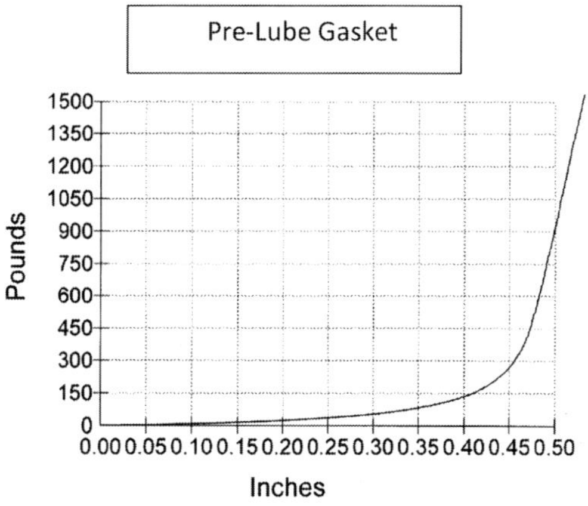

From the data on the table and graph shown in **Fig. 25**, 1st Force represents a minimum force that will be applied by the gasket at a maximum joint gap and 2nd Force represents at nominal gap.

A minimum 1st Force is 86.934 lb, which represents 62.3 psi; a nominal force of 140.5 lb represents 100.7 psi.

Variables exist such as stress relaxation on the rubber, joint opening, and so on; 62.3 psi can be reduced roughly by 25 %, which will provide around 47 psi of power. The safety factor will be three times that of the 15-psi hydrostatic requirements.

Summary

Simply put, precast box culverts are of superior quality, eliminate the danger of open trenches, and allow immediate backfill. These assets abolish the inconvenience of disruptive road closures.

Although gasketed box culverts require a considerable amount of investment, they are nevertheless capable of being manufactured precisely and are able to be used for special projects, which will make a quick return. Currently, several box section manufacturers have begun assembling their structures with a single offset joint. Simple gasketed box sections can be used to solve an assortment of project challenges. Displaying an increase in success, many consulting engineers and owners are favorably examining gasketed precast concrete box sections. These are easier to install, cost-effective, and provide high-performance results that are often superior to other materials or methods. The single offset joint, along with radius corners, is multifunctional. It allows the use of any type of profile gasket, with pre-lubricated gaskets most commonly utilized. Super seal gaskets allow for easy gasket installation and homing of the sections in the field. This type of joint will also accommodate butyl sealant or a joint wrap (or both), if specified by the project engineer.

The sealing method of joints for the box section (such as sealant, external wrap, or expanding water stop sealant) depends upon the installer's workmanship, the conditions, and the climate. The joints connecting the precast sections are of the greatest concern. It is recommended that ASTM C1677 specifications should be followed for future projects that use precast concrete box sections.

References

[1] "Concrete Box Culvert, Historical Overview," *Small Structures on Maryland's Roadways*, Maryland State Highway Administration, Baltimore, MD, 1997, http://sha.md.gov/OPPEN/Ch_3-2-2.pdf

[2] ASTM Committee C13 on Concrete Pipe, ASTM International, West Conshohocken, PA, www.astm.org

STP 1601, 2017 / available online at www.astm.org / doi: 10.1520/STP160120160116

Shawn R. Coombs[1] and John Kurdziel[2]

Calculation Variations Between the Indirect and Direct Design Methods

Citation

Coombs, S. R. and Kurdziel, J., "Calculation Variations Between the Indirect and Direct Design Methods," *Concrete Pipe and Box Culverts, ASTM STP1601*, J. J. Meyer and J. Beakley, Eds., ASTM International, West Conshohocken, PA, 2017, pp. 135–147, http://dx.doi.org/10.1520/STP160120160116[3]

ABSTRACT

Recently produced South Carolina Department of Transportation (SCDOT) reinforced concrete pipe (RCP) design standards capped the use of the indirect design method to RCP with diameters less than 36 in. in diameter and limited the use of the direct design method to RCP greater than or equal to 36 in. in diameter. In the original Marston and Spangler research done in the early twentieth century on concrete tiles with diameters up to 36 in. flexure was the governing failure mode. In the latter half of the twentieth century, the American Concrete Pipe Association (ACPA) initiated a long-range study to determine the failure modes of large diameter RCP. Out of the long-range research came the standard installation direct design (SIDD) method, which adds crack control and radial and diagonal tension to flexure as potential RCP failure modes. Today, because of the simplicity of the indirect design method and the popularity of the three-edge bearing test, many designers and end users are choosing the indirect design method for large diameter pipe designs, not realizing that this design method does not directly address all of the RCP modes of failure. This paper looks at the direct and indirect design methods and outlines potential issues that may arise by designing RCP using the indirect or direct design methods independent of each other.

Manuscript received August 30, 2016; accepted for publication January 3, 2017.
[1]Foltz Concrete Pipe and Precast, 11875 North Carolina Highway 150, Winston-Salem, NC 27127
[2]Advanced Drainage Systems, Inc., 4640 Trueman Blvd., Hilliard, OH 43026
[3]ASTM Symposium on *Concrete Pipe and Box Culverts* on December 7, 2016 in Orlando, Florida.

Keywords
direct design, indirect design, concrete pipe design, reinforced concrete pipe (RCP), RCP design, indirect, direct, design

Introduction

Since the late nineteenth century, concrete pipe has been used for drainage and sewer applications with relatively good success. In the first half of the twentieth century, Anson Marston and M. G. Spangler conducted research that led to the development of earth load and load factors used in the creation of the indirect and direct design methods for concrete pipe. Then, in the latter half of the twentieth century, the American Concrete Pipe Association (ACPA) initiated a long-range research project using enhanced techniques in structural analysis that resulted in the creation of the American Society of Civil Engineers' (ASCE) *Standard Practice for Direct Design of Buried Concrete Pipe* (SIDD).

Because both the indirect and direct design methods have been available for years, inexperienced designers occasionally fail to understand the applicability of each design methodology and fail to accommodate for each method's shortcomings. In some cases, concrete pipes can be overdesigned and, in other instances, they are underdesigned—not meeting strength or service limits requirements.

The South Carolina Department of Transportation (SCDOT) and the Nebraska Department of Roads (NDOR) are among a handful of state agencies in the twenty-first century that have identified potential inconsistencies between the indirect and direct design standards and have taken steps to either research or implement change (or both) to ensure strength and service limits are met, ensuring the performance of their installed RCP.

In an effort to more fully understand the differences between the indirect and direct design method, Foltz Concrete Pipe and Precast (Foltz), a Division of Advanced Drainage Systems, Inc. (ADS), conducted a literature review, design analysis, and field verification process looking into the differences between the indirect and direct design methods. Key results of this effort are listed in this paper.

Brief History of Indirect and Direct Design Method Development

In 1910 at what is now Iowa State University, Anson Marston began his theoretical study of nonreinforced concrete tiles used in farm drainage and sewer lines. The research revealed cracking in many of the nonreinforced rigid pipes greater than 15 in. in diameter. His research also revealed the need to develop methods to establish the load, its distribution on pipe, and to determine the supporting strength of pipe. In 1913, Marston and A. O. Anderson published their findings in "The Theory of Loads on Pipes in Ditches and Tests of Cement and Clay Drain Tile and Sewer Pipe" [1], and in 1930, Marston published "The Theory of External Loads on Closed Conduits in the Light of the Latest Experiments" [2]. In the 1930s, M. G. Spangler related an installed rigid pipe's ability to withstand Marston's earth loads to a smaller three-edge bearing load that produced the same invert moment. Spangler's comprehensive paper was entitled "The Supporting Strength of Rigid Pipe Culverts" [3]. Spangler's original "load factor" used in this comparison is known today as the "bedding factor." Spangler continued his work on embankment and negative projecting installation conditions and published his findings in "Field Measurements of the Settlement Ratios of Various Highway Culverts" in 1950 [4] and in "A Theory on Loads on Negative Projecting Conduits" [5] in 1951.

The indirect design method is the result of the ground-breaking procedures developed by Marston that Spangler used to calculate bedding factors for pipe that related the total field load applied to the pipe to the load applied in the three-edge bearing test. Today, the indirect design method is widely accepted by industry as a simple way to relate manufacturing proof of performance testing directly to field pipe performance. The performance criteria for the three-edge bearing test requires pipe to withstand laboratory loads for the 0.01 in. load condition and an ultimate strength under essentially two point loads without consideration of any lateral support. Resulting moments, thrusts, and shears from the earth pressure and their distribution around the rigid pipe are not taken into consideration in the three-edge bearing test. These impacts are empirically estimated with the use of bedding factors in the indirect design method. Some conditions, such as diagonal tension, radial tension, and field crack control, cannot be properly accommodated in the three-edge bearing test, so most of the correlation between the indirect design method and the three-edge bearing test is done for flexural moments.

For centuries, the direct design method has also been used to design buried rigid pipe. The direct design of buried pipe requires the determination of total load on the pipe and the distribution of earth pressure around the pipe. Total load is usually calculated using methods developed by Marston and Spangler, and the distribution of earth pressure was traditionally determined using either the Paris uniform distribution or Olander's radial earth pressure distribution. Once load and pressure distribution have been defined, the next step is to determine the structural effect of these loads in the pipe wall. The structural effects are defined in terms of bending moments, thrusts, and shears at all points in the pipe ring.

Because of advances in techniques for structural analysis and the identification of shortcomings in the indirect design and traditional direct design practices, a long-range research program to develop a new direct design procedure for concrete pipe in the installed condition was initiated by the ACPA. The result of this long-range research program was implemented through the creation of the ASCE's specification, ASCE 15-98, *Standard Practice for Direct Design of Buried Concrete Pipe Using Standard Installations* (also known as SIDD).

The SIDD design is based on limit states that provide assurance that the pipe will have adequate strength and serviceability. Standard installations (Type 1, 2, 3, and 4) are used in the design, and the enhanced Heger soil pressure distribution model with vertical and horizontal arching factors was added to the design [6]. The installation types and the Heger soil pressure distribution are intended to represent current installation practices found in the market today. The SIDD method was incorporated into the Pipe Culvert Analysis and Reinforcement (PIPECAR) design program.

Indirect versus Direct Design Calculation

"In order to take advantage of advances in knowledge about the behavior of structures, in the 1970s and '80s, ACPA spearheaded the development of new standards for concrete pipe and box sections. They also initiated a major long-range research program to serve as a basis for new, more direct design approaches for buried concrete pipe based on the behavior of pipe in installed condition" [7]. To help explain the benefits of the direct design method, **Fig. 1** was developed. "Figure 1 shows how the amount of inside reinforcement required at the pipe invert typically varies with the height of earth cover above the top of the pipe (fill height) for pipe without stirrups" [8]. "The figure shows that flexural strength is the governing design criteria for the initial and largest portion of the range of fill heights that can be supported using concrete pipe without stirrups" [8]. When shear governs, "there is an additional small range of fill heights that can be supported by pipe without stirrups by increasing the inner circumferential reinforcement substantially beyond the increases that would be required for flexural strength or for crack control in order

FIG. 1 Plot of required inside reinforcing area versus design height of earth cover for typical design with surface wheel loads.

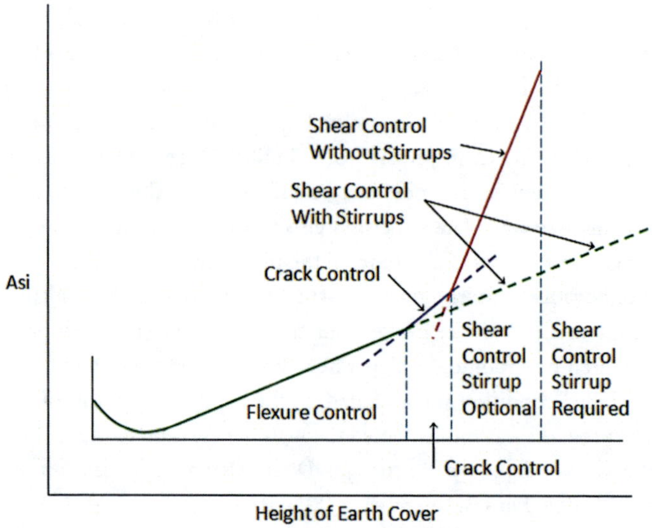

to meet the requirements for shear strength without the use of stirrups" [9]. "However, in the indirect method, the earth pressures and their distribution around the pipe and the resulting moments, thrusts, and shears in the pipe are not calculated. Instead, procedures developed by Marston-Spangler..., are used to calculate the bedding factors for pipe, which relate the total field load applied to the pipe to the load applied in the three-edge bearing test" [7].

Indirect design looks at flexural failure and the region where service cracks and shear govern are missed. The lack of service crack and shear analysis is particularly evident when comparing indirect versus direct designed pipe larger than 42 in. in deep fill (Class IV and Class V) with C-wall configurations. In an effort to highlight the critical nature of incorporating shear into design, the fill heights listed in **Table 1** were developed using the Federal Highway Administration's accepted direct design

TABLE 1 Indirect versus direct fill height chart.

Class III RCP																	
B-Wall									C-Wall								
	Type 1		Type 2		Type 3		Type 4			Type 1		Type 2		Type 3		Type 4	
Diam. (in.)	ACPA Fill (ft)	FHWA Fill (ft)	ACPA Fill (ft)	FHWA Fill (ft)	ACPA Fill (ft)	FHWA Fill (ft)	ACPA Fill (ft)	FHWA Fill (ft)	Diam. (in.)	ACPA Fill (ft)	FHWA Fill (ft)	ACPA Fill (ft)	FHWA Fill (ft)	ACPA Fill (ft)	FHWA Fill (ft)	ACPA Fill (ft)	FHWA Fill (ft)
36	23	21	17	13	13	9	9	4	36	23	9	17	5	13	5	9	5
42	23	21	17	13	13	9	9	NA	42	23	12	17	6	13	NA	9	NA
48	23	21	17	13	13	9	9	4	48	23	14	17	8	13	NA	9	NA
54	22	21	17	14	13	10	9	5	54	22	21	17	10	13	7	9	NA
60	22	21	17	15	13	11	9	6	60	22	18	17	11	13	7	9	NA
72	22	20	17	15	13	11	9	7	72	22	20	17	13	13	9	9	5
84	21	20	16	15	12	11	9	7	84	21	20	16	15	12	11	9	7
96	21	19	16	15	12	11	8	7	96	21	19	16	15	12	11	8	7
108	21	19	16	14	12	10	8	6	108	21	19	16	14	12	10	8	6
Class IV RCP																	
B-Wall									C-Wall								
	Type 1		Type 2		Type 3		Type 4			Type 1		Type 2		Type 3		Type 4	
Diam. (in.)	ACPA Fill (ft)	FHWA Fill (ft)	ACPA Fill (ft)	FHWA Fill (ft)	ACPA Fill (ft)	FHWA Fill (ft)	ACPA Fill (ft)	FHWA Fill (ft)	Diam. (in.)	ACPA Fill (ft)	FHWA Fill (ft)	ACPA Fill (ft)	FHWA Fill (ft)	ACPA Fill (ft)	FHWA Fill (ft)	ACPA Fill (ft)	FHWA Fill (ft)
36	34	33	26	25	20	20	14	13	36	34	21	26	13	20	9	14	5
42	34	33	26	25	20	19	14	13	42	34	24	26	15	20	11	14	7
48	34	33	26	25	20	19	14	13	48	34	27	26	18	20	13	14	8
54	34	33	26	24	20	19	14	13	54	34	33	26	20	20	15	14	10
60	34	32	26	24	20	18	14	13	60	34	32	26	22	20	17	14	11
72	33	32	25	23	20	17	14	12	72	33	32	25	22	20	17	14	11
84	SD	-	SD	-	SD	-	SD	-	84	33	31	25	24	19	17	14	12
Class V RCP																	
B-Wall									C-Wall								
	Type 1		Type 2		Type 3		Type 4			Type 1		Type 2		Type 3		Type 4	
Diam. (in.)	ACPA Fill (ft)	FHWA Fill (ft)	ACPA Fill (ft)	FHWA Fill (ft)	ACPA Fill (ft)	FHWA Fill (ft)	ACPA Fill (ft)	FHWA Fill (ft)	Diam. (in.)	ACPA Fill (ft)	FHWA Fill (ft)	ACPA Fill (ft)	FHWA Fill (ft)	ACPA Fill (ft)	FHWA Fill (ft)	ACPA Fill (ft)	FHWA Fill (ft)
36	52	51	40	39	31	31	22	21	36	52	44	40	28	31	22	22	12
42	52	49	40	33	31	26	22	18	42	52	47	40	31	31	24	22	17
48	52	47	40	31	31	25	22	17	48	52	48	40	31	31	24	22	17
54	SD	-	SD	-	SD	-	SD	-	54	52	44	40	29	31	23	22	16
60	SD	-	SD	-	SD	-	SD	-	60	51	44	40	29	31	22	22	16
72	SD	-	SD	-	SD	-	SD	-	72	51	40	39	26	30	20	22	14

method in comparison to the ACPA's published indirect design values. The direct design fill heights were calculated using PIPECAR 4.0's direct design module. Inputs for these calculations are listed in the Appendix in Charts 1a, 1b, and 1c. The indirect design fill heights were taken from the ACPA's published fill height tables developed using the indirect design method in accordance with Section 12.10.4.3 of the American Association of State Highway and Transportation Officials Load and Resistance Factor Design (AASHTO LRFD) *Bridge Design Specification* [8,10].

Highlighted in **Table 1** are the variations in acceptable fill heights between the ACPA's LRFD indirect designed fill heights and the AASHTO LRFD direct designed fill heights. Fill heights were calculated using pipe diameters, wall thickness, and steel areas referenced in ASTM C76, *Standard Specification for Reinforced Concrete Culvert, Storm Drain, and Sewer Pipe* [11], which only provides steel areas to meet various three-edge bearing strengths, not a D-load design [12]. It is this disconnect between the laboratory three-edge bearing strength and the in-field performance that creates much of the discontinuity with the indirect design method and direct design method. It is clear that the variation between the indirect and direct designs for B-walls appears to be minimal, but variation in allowable fill heights for C-wall, particularly in Class IV and V pipe, are significant. Values for diameters less than 36 in. were excluded from **Table 1** because the direct design was created primarily for larger diameter pipes and is overly conservative when designing pipes with diameters less than 36 in.

Variance among fill heights produced by the two design methods suggests that correlation between the two designs should be undertaken.

SCDOT'S Indirect versus Direct Comparison

In April 2009, the SCDOT produced reinforced concrete pipe (RCP) design standards that capped the use of the indirect design method to RCP less than 36 in. in diameter and limited the use of the direct design method to RCP greater than or equal to 36 in. in diameter [13]. The creation of an updated fill height chart was the result of an in-house review of the indirect and direct design methods as well as a massive effort by the SCDOT to correlate their installation standards to their published fill heights.

In September 2012, the SCDOT modified their published fill heights to align with the AASHTO standards, which allow the use of both the direct and indirect design methods [14].

The 2009 and 2012 fill heights are listed in **Table 2**. It must be noted that fill heights generated by the SCDOT are specific to the DOT's design criteria and installation requirements. It should also be noted that, in reverting back to the indirect design, 36-in. and 42-in. diameter pipes were restricted to B-wall designs.

NDOR Indirect versus Direct Comparison

In June 2006, the NDOR produced the results of a research project they conducted entitled "Behavior and Design of Buried Concrete Pipes" [15]. Prior to the research,

TABLE 2 SCDOT fill heights by pipe class and year published.

RCP Diam. (in.)		Class III Apr. 2009	Class IV Apr. 2009	Class V Apr. 2009		Class III Sept. 2012	Class IV Sept. 2012	Class V Sept. 2012
12	Indirect Design	16	25	30	Indirect Design	16	25	30
15		16	25	30		16	25	30
18		16	25	30		16	25	30
24		16	25	30		16	25	30
30		16	25	30		16	25	30
36	Direct Design	NA	13	28		16 (B)	25 (B)	30
42		NA	16	27		16 (B)	25 (B)	30
48		10	17	27		16	25	30
54		11	17	27		16	25	30
60		12	NA	26		15	25	30
66		13	NA	26		15	25	30
72		13	NA	26		15	24	30
78		12	NA	18		15	24	CUST
84		12	NA	18		15	24	CUST
90		14	CUST	CUST		15	CUST	CUST
96		14	CUST	CUST		14	CUST	CUST
108		14	CUST	CUST		14	CUST	CUST
120		CUST	CUST	CUST		CUST	CUST	CUST

NDOR used with reliability both the indirect and direct design methods for designing concrete pipe, but as a result of advancements in manufacturing and construction, they initiated the research project to verify if their practices were economic and up-to-date.

A portion of the NDOR report appears in **Table 3**, which lists fill heights for ASTM C76 Class III, IV, and V pipes using NDOR's standard design practice, the AASHTO standard, the AASHTO LRFD, and ACPA indirect design methods. After closely reviewing the design methods, it was determined that the NDOR's standard practices of using the direct design to determine fill heights shown in **Table 3**, based on existing pipes they had in inventory, was appropriate, and that the AASHTO LRFD values (shaded) were recommended.

It was also emphasized that the direct design allowed for greater variation in design than did the indirect design, which limited pipes to steel areas listed in ASTM C76. In many cases, the direct design would allow for a more economic pipe design.

Indirect versus Direct Design Tables

As illustrated by the variability in calculated fill heights listed in **Tables 1**, **2**, and **3**, it appears that prudent designers will want to correlate results from the indirect and direct designs to ensure proper field performance and to verify they have achieved an optimal design while meeting required strength factors.

TABLE 3 NDOR fill height table comparison [15].

RCP Diam. (in.)	Class III Fill Height (ft)				Class IV Fill Height (ft)				Class IV Fill Height (ft)			
	NDOR	STD	LRFD	ID	NDOR	STD	LRFD	ID	NDOR	STD	LRFD	ID
15	12	12	13	14	15	15	16	22	21	21	22	33
18	12	12	13	15	17	17	18	22	24	24	25	34
21	13	13	13	15	19	19	20	22	26	26	27	34
24	13	13	12	15	19	19	20	22	26	26	27	34
27	13	13	13	14	17	17	17	22	26	26	27	34
30	12	12	12	14	14	14	15	22	25	25	25	33
36	10	10	11	14	16	16	17	22	24	24	25	33
42	10	10	11	14	15	15	16	22	23	23	24	33
48	10	10	11	14	14	15	15	21	23	23	24	33
54	10	10	11	14	14	15	15	21	-	-	-	-
60	9	10	10	14	14	15	16	21	-	-	-	-
66	9	10	10	14	14	16	16	21	-	-	-	-
72	9	10	10	13	14	16	16	21	-	-	-	-
78	9	10	11	13	-	-	-	-	-	-	-	
84	9	10	10	13	-	-	-	-	-	-	-	-
90	9	10	11	13	-	-	-	-	-	-	-	-
96	9	10	11	13	-	-	-	-	-	-	-	-
102	10	11	11	-	-	-	-	-	-	-	-	-
108	10	11	11	-	-	-	-	-	-	-	-	-

Note: NDOR = NDOR standard design practice; STD = AASHTO STD, LRFD = AASHTO LRFD, and ID = ACPA indirect design.

It should also be noted that ASTM C76 is a manufacturing and purchase specification only, and does not include requirements for bedding, backfill, or the relationship between field load condition and the strength classification of pipe. D-load testing is based on the maximum allowable moment, where the direct design reviews flexure, crack control, radial tension, and diagonal tension failure.

Field Verification

After studying literature and identifying significant variability in fill height values between the indirect to direct design methods, a brief field investigation was conducted to see if field issues correlated calculated findings.

It took three stops along a randomly selected highway to identify a 42-in. ASTM C76, Class III, C-wall pipe that was installed under approximately 22 ft of fill with flexural cracks (**Fig. 2**). Per ACPA fill heights, if the pipe was installed using a Type 1 installation, it would have met service load criteria. However, the pipe was cracked in the invert at 0.12 in., (**Fig. 3**) and **Fig. 4** had dual flexure cracks in the crown, the larger of the two measuring at 0.035 in. In the pipe with 15 ft of cover,

FIG. 2 Manufacturing data found inside the in-field inspected ASTM C76, Class III, 42-in. pipe.

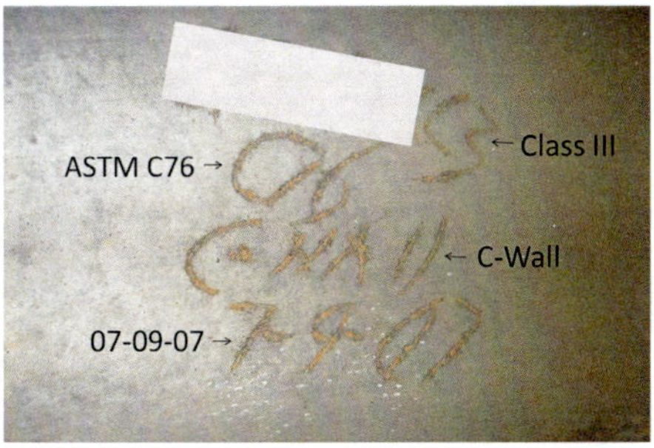

the cracks in the pipe's crown were just under 0.01 in. In this case, had the more conservative fill height been used based on values listed in **Table 1**, the maximum allowable fill would have been appropriate for the application and might not have resulted in flexural cracking. It is important to note that because the referenced 42-in. diameter pipe was under a public highway, it was not possible to identify the installation parameters or to evaluate the pipe's properties. This example simply highlights that field flexural cracks existed for a pipe diameter where the indirect and direct design have significant variability.

This small field sample highlights that field issues exist but does not have enough data to evaluate the validity of the indirect or direct design method. After this initial site visit, it was determined that a full field study of buried pipe would be required, which fell outside the limits of this study.

FIG. 3 In-field inspected ASTM C76, Class III, 42-in. RCP with approximately 22 ft of cover and a 0.12-in. crack in the pipe's invert.

FIG. 4 In-field inspected ASTM C76, Class III, 42-in. RCP with approximately 22 ft of cover and a 0.035-in. crack in the pipe's crown.

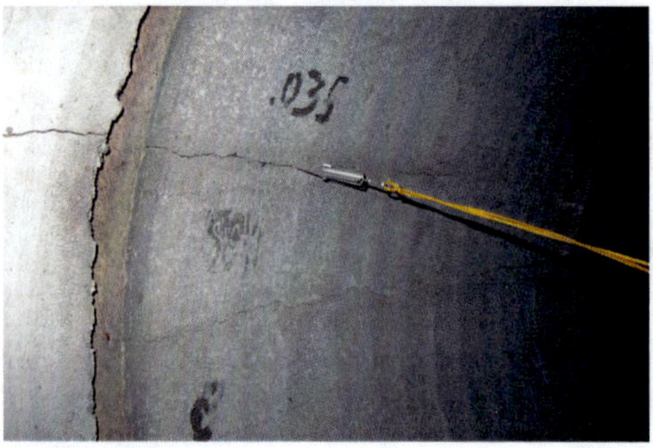

Conclusions

Maximum allowable fill heights can vary when comparing indirect and direct designs for pipe with identical material properties, which is confusing to design engineers and leads to lack of confidence in the design methodologies. It is also easy to confuse D-loads and the three-edge bearing test as a design method. The three-edge bearing test is based on the maximum allowable moment and is a manufacturing performance test, not a design method. The direct design method provides uniform load factors and allows engineers to design pipe to specific in-field conditions. Variability between the indirect and direct design values was greatest for ASTM C76 Class IV and V C-wall pipes with diameters larger than 36 in.

Both the direct and indirect design methods are approved by AASHTO but, as presented in this paper, the variations in results from the two systems suggests a study should be done to identify and correlate variability between the two design methods.

The field inspection did not disprove deficiencies in the indirect or direct design, but it did highlight that a more conservative design approach may be warranted.

References

[1] Marston, A. and Anderson, A. O., "The Theory of Loads on Pipes in Ditches and Tests of Cement and Clay Drain Tile and Sewer Pipe," *Iowa State College of Agriculture and Mechanic Arts, Bulletin No. 31*, Vol. 11, No. 5, 1913.

[2] Marston, A., "The Theory of External Loads on Closed Conduits in the Light of the Latest Experiments," *Iowa State College of Agriculture and Mechanic Arts, Bulletin No. 96*, Vol. 28, No. 38, 1930.

[3] Spangler, M. G., "The Supporting Strength of Rigid Pipe Culverts," Bulletin 112, Iowa Engineering Experiment Station, Ames, IA, 1933.

[4] Spangler, M. G., "Field Measurements of the Settlement Ratios of Various Highway Culverts," Bulletin 170, Iowa Engineering Experiment Station, Ames, IA, 1950.

[5] Spangler, M. G., "A Theory on Loads on Negative Projecting Conduits," presented at the *Thirtieth Annual Meeting of the Highway Research Board*, Washington, DC, January 9–12, 1951, Highway Research Board, Washington, DC, 1951.

[6] Heger, F. J. and McGrath, T. J., *Design Method for Reinforced Concrete Pipe and Box Sections*, American Concrete Pipe Association, Irving, TX, 1982.

[7] American Concrete Pipe Association, *Concrete Pipe Technology Handbook*, ACPA, Irving, TX, 2001.

[8] American Concrete Pipe Association, *LRFD Fill Height Tables for Concrete Pipe*, ACPA, Irving, TX, 2013.

[9] ASCE 15-98, *Standard Practice for Direct Design of Buried Precast Concrete Pipe Using Standard Installations (SIDD)*, American Society of Civil Engineers, New York, 2001.

[10] AASHTO, *LRFD Bridge Design Specifications*, 6th ed., American Association of State Highway and Transportation Officials, Washington, DC, 2013.

[11] ASTM C76, *Standard Specification for Reinforced Concrete Culvert, Storm Drain, and Sewer Pipe*, ASTM International, West Conshohocken, PA, 2014, www.astm.org

[12] ASTM C497, *Standard Test Methods for Concrete Pipe, Manhole Sections, or Tile*, ASTM International, West Conshohocken, PA, 2016, www.astm.org

[13] South Carolina Department of Transportation, Standard Drawing 714-205-00, Pipe Culverts (Rigid Reinforced Concrete Pipe (RCP) Details and Fill Height), April 2, 2009.

[14] South Carolina Department of Transportation, Standard Drawing 714-205-00, Pipe Culverts (Rigid Reinforced Concrete Pipe (RCP) Details and Fill Height), September 26, 2012.

[15] Erdogmus, E. and Tadros, M. K., "Behavior and Design of Buried Concrete Pipes," *Nebraska Department of Roads Research Reports*, Paper 54, NDOR, Lincoln, NE, 2006.

Appendix

CHART 1A Direct Design Inputs for Creation of Table 1. Direct Design Fill Heights

Key notes, definitions, and PIPECAR 4.0 inputs used to calculate direct design fill heights listed in the indirect versus direct design fill height comparison chart.

Key Notes and Definitions	**1) Whenever changing pipe diameter or from B-Wall (B) to C-Wall (C), a new file must be created**. Otherwise, the defaults will not change and the data will not be repeatable. This is also the case when pipe diameters are modified. 2) PIPECAR Version 4.0 was used, along with the values listed as follows, to create the subsequent calculations.

	3) NA = No acceptable design was found when highway loads are applied to the design. Earth load exceeded the materials' performance properties in deeper fills and live load exceeded the materials' performance properties in shallow cover.
	4) CCP = The program produced an error message stating, "Design not possible due to excessive concrete compression." A double cage helped solve this issue.
	5) DCR = Double cage required for design to work.
PIPECAR 4.0 - Inputs, Page 1:	
Pipe Shape	Pipe shape circular
	Pipe wall thickness followed ASTM C76 for B-wall default value in program; then, for C-wall, modify from default value.
Materials Properties	Steel reinforcing yield strength 65.0 ksi
	Reinforcing Type 2
	Design concrete strength: **6.0 ksi Changed from default value of 5.0 ksi.**
	Concrete density: 150 pcf
S.P.D.	Soil pressure distribution: Heger pressure distribution

CHART 1B Direct Design Inputs for Creation of Table 1. Direct Design Fill Heights – Continued.

PIPECAR 4.0 Inputs, Page 2:	Design code: AASHTO LRFD
Design Code Load Factors	Dead load moment and shear: Load factor 1.3, modifier 1.05
	Lead load thrust: Load factor 1.0, modifier 1.0
	Live load moment and Shear: Load factor 1.75, modifier 1.0
	Live load thrust: Load factor 1.0, modifier 1.0
	Internal Pressure Thrust: Factor 1.00, modifier 1.0
Strength Reduction Factors	Flexure: 1.0
	Diagonal tension: 0.9
	Radial tension: 0.9
	Limiting crack width factor: 1.0; changed from default value of 0.9
Process Factors	Radial tension process factor: 1.00
	Shear process factor: 1.0
Installation Condition SIDD Soil Pressures	Installation type: Used Types 1, 2, 3, and 4
	Height of earth fill: Varied based on ACPA's LRFD fill height tables for concrete pipe
	Do you wish to change the defaults? No
	Vertical arching factor: Left as default, which varied based on installation type.
	Horizontal arching factor: Left as default, which varied based on installation type.
PIPECAR 4.0 - Inputs, Page 3:	
PIPECAR 4.0 Inputs, Page 3:	Soil unit weight: 120 pcf

Soil & Fluid Load Data:	Depth of fluid: Equaled pipe's inside diameter Fluid unit weight: 62.4 pcf Pressure head: 0 ft
Highway Live Load Parameters	Live load data: Highway Single axle load: 32 kips Load per axle of double axle load: 25 kips Tire footprint length: 10 in. Tire footprint width: 20 in. Lane load: 64 psf Direction of traffic: Across pipe Impact factor: Design Code 1.33

CHART 1C Direct Design Inputs for Creation of Table 1. Direct Design Fill Heights - Continued

PIPECAR 4.0 Inputs, Page 4:	Reinforcing cage type: We used double circular when ASTM C76 showed values in the Asi and Aso position and single circular when ASTM C76 only used Asi.
PIPECAR 4.0 Inputs, Page 4: R.C.T. Concrete Cover	Inside face: 1.00 in. Outside face: 1.00 in.
Reinforcing Diameter	Inside reinforcing diameter, Asi: Used default values, which varied. Outside reinforcing diameter, Aso: Used default values, which varied.
Maximum Reinforcing Spacing	Inside reinforcing spacing, Asi: 2.00 in. Changed from default value of 4.00 in. Outside reinforcing spacing, Aso: 2.00 in. Changed from default value of 4.00 in.
PIPECAR 4.0 - Inputs, Stirrup Reinforcing Routine:	
PIPECAR 4.0 Inputs, Stirrup Reinforcing Routine: Section - 1	Developable stirrup yield stress: 60 ksi. Changed from default value of 40 ksi. Stirrup spacing (maximum = 0.75 fd): 1.00 in. Required steel area for stirrups: 0.052 (in.^{2}/ft)/line Required number of lines: 15 Stirrups centered on: Invert